Keeping Women in Science

Keeping Women in Science

Kate White

MELBOURNE
UNIVERSITY
PRESS

MELBOURNE UNIVERSITY PRESS
An imprint of Melbourne University Publishing Limited
11–15 Argyle Place South, Carlton, Victoria 3053, Australia
mup-info@unimelb.edu.au
www.mup.com.au

First published 2014
Text © Kate White, 2014
Design and typography © Melbourne University Publishing Limited, 2014

This book is copyright. Apart from any use permitted under the *Copyright Act 1968* and subsequent amendments, no part may be reproduced, stored in a retrieval system or transmitted by any means or process whatsoever without the prior written permission of the publishers.

Every attempt has been made to locate the copyright holders for material quoted in this book. Any person or organisation that may have been overlooked or misattributed may contact the publisher.

Text design by Phil Campbell
Cover design by Phil Campbell
Typeset by J&M Typesetting
Printed in Australia by Opus Group

National Library of Australia Cataloguing-in-Publication entry

White, Kate, 1949 August 3– author.
Keeping women in science Dr Kate White.

9780522867015 (paperback)
9780522867022 (ebook)

Includes bibliographical references and index.
Women in science—Australia.
Women in the professions—Australia.
Work–life balance—Australia.

305.4350994

Contents

1	Introduction	1
2	Literature Review	12
3	Research Design	22
4	Analysis of the Institute's Workforce	32
5	Job Satisfaction	43
6	Building Career Paths in Science Research	54
7	Networking, Mobility and Mentoring	69
8	Gender and Career Paths	93
9	Generational Change in Science Research	119
10	Keeping Women in Science Research	141
11	New Ways of Doing Science	156

Afterword	173
Appendix	176
Acknowledgements	178
References	179
Index	188

CHAPTER 1
Introduction

Why is it so important to keep women in science research? The Australian Government's Innovation Agenda asserts that an internationally competitive economy begins with an internationally competitive innovation system, and that begins with internationally competitive universities (Po*wering Ideas* 2009). The nation cannot afford to invest in educating women to PhD level and then see them exit the field because they experience barriers and/or are trying to juggle family and career.

Keeping women in science has now become a focus of both national and international policymakers and higher education (HE) leaders. Women scientists leave the profession in greater proportions than men and are under-represented in leadership roles. A recent UK House of Commons committee report asserted that one compelling reason to tackle the problem of the under-representation of women at senior levels in Science, Technology, Engineering and Mathematics (STEM) 'is that the UK economy needs more STEM workers and we cannot meet the demand without increasing the numbers of women in STEM' (House of Commons (HoC) 2014, p. 3), while an Australian inquiry into medical health research called for universities and research institutions to implement gender equity policies that would attract more women to science research (McKeon 2013, p. 138).

Purpose and Context

This book focuses on a large Melbourne research institute as a case study of how to improve institutional practices that can support career progression, particularly for women in science. It identifies current barriers to career progression, including gender and equity and disparity of outcomes, and recommends strategies that can be implemented to address these barriers.

The broad context for the study is the 2009 report by Professor Sharon Bell on women in science in Australia that demonstrated the disadvantage of women in career progression. She asserted that: '… when we document "attrition" we are mapping accumulated disappointment, frustration and unrealised expectations, impacting significantly on individuals'. In response, Bell contended that 'it is our responsibility to change the professional world our young scientists are entering' (Bell 2009, p. 57; see also Williams 2010).

Bell (2009) argued that in order to move forward it was necessary to: identify barriers and the accumulation of disadvantage; accommodate and support flexible and non-traditional career paths, including transition support programs; recognise and reward professional excellence relative to opportunity; identify biased and subjective evaluation criteria that impede progress; and provide access to role models, mentoring and professional networks. Since the release of Bell's report, a summit at Parliament House Canberra on 11 April 2011 resulted in commitments to undertake real change to address women's participation in science and engineering.

Scientific careers have historically represented—and continue to represent—a challenge for women, characterised by minority status, gendered division of labour both horizontally and vertically, and low representation in gatekeeping and decision-making positions (genSET 2010; Caprile 2012). genSET (2010, p. 22) argues that there has been a significant change in scientific research. Whereas historically the three fundamental purposes of research in science were: to disseminate new information so that others could learn from it, to set a foundation for other scientists to build on or repeat the studies with additional observations or experiments and to prove to interested parties the scientist's findings. Now, publication is the scientist's lifeline and has become the main goal. This enormous change in emphasis has damaged the practice of science, has transformed the motivation of researchers, changed the way results are presented, and reduced the accuracy and accessibility of the scientific literature (genSET 2010). This narrow focus on publication in high-impact scientific journals is propelled by a funding model, at least in Australia, that can limit participation in scientific research.

The Funding Model for Science Research

Most national funding for science research in Australia comes from the National Health and Medical Research Council (NHMRC), as discussed later in the chapter, and is therefore hugely important in the career progression of research scientists. Without ongoing external funding it is impossible to build a successful career. And yet the effectiveness of this funding model has been questioned. Graves et al. (2011) obtained the category and summary scores for all project grant applications considered in the 2009 funding round. The NHMRC committed 50.3 per cent of its $714 million annual budget to the project grants scheme that year. Applications were between seventy and 120 pages, including a nine-page research plan.

Graves et al. argued that the assessment of grant proposals was costly and subject to a high degree of randomness owing to variation in panel members' assessments. The total cost per proposal was $17,744 with around 85 per cent of that cost incurred by applicants. The median estimate of twenty-two days preparing a grant multiplied by the 2,983 grant proposals submitted showed that preparing these applications used a total of 180 years of researchers' time.

Graves et al. (p. 3) asserted that the benefits of participating varied for applicants: those who scored in the top 9 per cent were always funded, while the next 29 per cent faced uncertainty and were not always funded, and the remaining 61 per cent faced 'certain outcomes of zero, as the variation among assessors was insufficient for them to ever score above the funding line'. Graves et al. therefore argued that for this last group the time invested was likely to be 'a deadweight loss other than some process utility from writing the grant and from participating in peer review. Applicants in this group might have benefitted more from doing something else with their time'. Graves et al.'s research therefore demonstrates that submitting a funding application could be both a costly and time-consuming process, and that the randomness of the assessment process meant that for the majority of applicants it was not an effective use of their time.

Funding is awarded on the basis of a researcher's track record, which entails securing grant funding for research and publishing research in high-impact journals. Assessing publication records by the number and impact of papers produced 'militates against career breaks or reduced working hours' (HoC 2014, p. 33). There is little flexibility in funding programs in Australia to cater for those applicants who want to work part-time or to spread their grant over a longer period (ACOLA 2012). Jacobson (2013) argues that the following six changes to the funding structure would help retain Australian women scientists who have had career disruptions: changing annual funding deadlines (NHMRC career development fellowship and project grant deadlines are only a few weeks apart each year and disadvantage women on maternity leave); maternity leave support for project grants; increasing grants from three to five years (three-year grants put pressure on researchers to build their track record in a short time and exacerbate the effect career disruptions have on funding for women); increasing the number of fellowships for senior postdocs; a clearer method of judging career disruptions for all components of a track record; and, finally, judging quality of publications over quantity (quantity of publications is often the main measure of track record and is biased against women who have reduced hours when returning to work after maternity leave). Herbert et al. (2014) supported the view that the timing of the funding cycle could be altered to better assist applicants in balancing work and family commitments.

In September 2011, the Australian government announced a review to consider how to optimise the future environment for health and medical research in this country in a fiscally sustainable manner. Two of the thirteen terms of reference were:

> Opportunities to improve coordination and leverage additional national and international support for Australian health and medical research through private sector support and philanthropy, and opportunities for more efficient use, administration and monitoring of investments and the

health and economic returns; including relevant comparisons internationally.

Strategies to attract, develop and retain a skilled research workforce which is capable of meeting future challenges and opportunities (McKeon 2013, p. 266).

Known as the McKeon Review, it reported its findings to the federal government in 2013. The review identified five key issues that needed to be addressed in relation to the research workforce. These were: career progression and salary barriers; career break impact on re-entry into the workforce; gender inequalities for both male and female researchers; lack of capacity to mentor young researches; and absence of viable career structures (McKeon 2013, p. 136). While the review made a key recommendation to put more researchers in the system through increasing flexibility around career breaks or part-time work, removing barriers to retention and funding for mentoring (McKeon Review 2013), the mechanism for implementation of this recommendation was not clearly expressed. The possible impact of the review's recommendations on the working lives of women and men in science research will be explored in later chapters.

Impact of the Funding Model on Science Research

The funding model for science research in Australia determines who is considered a successful researcher. The NHMRC grants and the fellowship system reward those who have prolific research output in high-impact scientific journals. This productivity requires scientists to work long hours and prioritise research above all else. It also assumes that the model of the successful scientist, discussed further in chapter 8, is that of a monastic male who has no responsibilities other than doing science; that is, no partner, no children, and no interests or activities away from the workplace. But even when men do have children this is not considered to impact greatly on their careers, because they are often not the primary care giver. Leadership positions are not associated with care obligations, but rather 'an unlimited commitment of time and space' to the organisation (von Alemann and Beaufays 2014, p. 1). Moir (2006, p. 8) argues that construction of the notion of a scientist as male results in gender-blind rhetoric that 'presents the role of professional scientists as virtually immutable given that science is taken to be the very male model of the rational pursuit of objective scientific knowledge that requires dedication to long hours of laboratory work. To try and change this role would be … almost tantamount to attempting to interfere with the very objective nature of science itself'.

The long hours work culture begins early in a research scientist's career. *Naturejobs* (blog 1 April 2011) reported that more than half of the postdocs surveyed worked at least fifty hours per week and a quarter worked at least sixty hours per week. Thus, at the outset of a science research career some postdocs are moving towards Hewlett and Buck Luce's (2006) concept of extreme jobs (involving a seventy-hour working week and other high-pressure characteristics).

The long hours culture is reinforced by those who head up laboratories. In the research institute considered in this study, most lab heads are men over fifty years of age who are the main breadwinners in their family. They have generally had the freedom to pursue their careers without the need to take into consideration the career path of their partner or to share the primary responsibility for childcare or care of older relatives. Charlesworth et al.'s (1989, p. 273) study of a high-profile Melbourne research institute, Walter and Eliza Hall Institute (WEHI), twenty-five years ago concluded that women scientists 'miss out' and added: 'The whole process of professionalisation (making one's run between the ages of twenty-five and thirty-five) is biased against them'.

With a new generation of science researchers about to move into leadership positions as the Baby Boomer generation approaches retirement, the model of a successful scientist promoted by the current funding model appears to be outdated. Younger male scientists need to consider the careers of their partners, as well as their own, and have a quite different view about balancing work and other responsibilities. In these more egalitarian relationships male scientists tend to split the workload at home with their partners (*Science Careers Blog*, 23 August 2012). Moreover, the partners in these egalitarian households also tend to be tenure-track scientists or professionals in other academic fields. Interestingly, some of these couples had decided not to have children. As the Executive Director and CEO of the American Association for Women in Science (AWIS 2012, p. 13) asserts, 'The real issue is that the academic workplace is still modelled on an ideal that no longer exists nor complements the realities of today's global workforce'.

The impact of the funding model on women scientists is even more pronounced. It is difficult for them to combine a science research career with caring responsibilities. This leads to significant attrition (Bell 2009), as discussed above. Many leave science research for industry or academia, and those who remain often combine research with clinical practice, rather than focusing on pure science research. It is yet to be determined if the recommendations of the McKeon Review (2013) will have a significant impact on this traditional gendered model of science excellence and in turn on the careers of women in science research.

A Case Study: The Florey Institute of Neuroscience and Mental Health

The Florey Institute of Neuroscience and Mental Health is one of the world's leading brain research centres, employing more than 500 staff and educating in excess of 100 post-graduate students each year. It has the largest neuroscience research team in Australia working across a variety of disease states such as stroke, epilepsy, Alzheimer's disease, Parkinson's disease, multiple sclerosis, Huntington's disease, motor neuron disease, traumatic brain and spinal cord injury, depression, schizophrenia, bipolar disorders, anxiety disorders and addiction. The institute is a world leader in central control of autonomic function, central regulation of fluid and electrolyte balance, neuropeptide chemistry and neurobiology, neural development, imaging technology, stroke, epilepsy, mental health and dementia.

State and federal governments, major philanthropic foundations and private benefactors have recognised the importance of neuroscience as the final frontier in medical research and have helped the institute and its partners build two state-of-the-art research facilities, one at the University of Melbourne in Parkville and the other at the Austin Hospital in Heidelberg costing in excess of $200 million.

The institute was originally known as the Howard Florey Research Institute, named after Lord Howard Florey, the Australian

Nobel laureate whose research work on penicillin continues to save millions of lives each year. It was established by an Act of State Parliament in 1971 as an independent medical research institute affiliated with the University of Melbourne.

It was initially a basic science medical research institute with a focus on physiology. However, Professor Fred Mendelsohn, the Institute's Director from 1997 to 2009, decided on a fundamental shift in direction from experimental physiology towards a major focus on neuroscience and neurological disease research. That led to the creation of the Florey Neuroscience Institutes (FNI) in April 2007, which was an amalgamation of the Howard Florey Institute and two other institutes based at the Austin Hospital in Heidelberg—the National Stroke Research Institute headed by Professor Geoff Donnan, and the Brain Research Institute with a strong focus on epilepsy and high field strength neuro-imaging headed by Dr Graham Jackson. Professor Donnan succeeded Professor Mendelsohn as Director of the FNI in January 2009.

In September 2012, the FNI amalgamated with the Mental Health Research Institute headed by Professor Colin Masters to create the Florey Institute of Neuroscience and Mental Health. These successive amalgamations have brought together four institutes in the brain and mind field as one significant entity on the global stage. As part of the process, the Florey Institute of Neuroscience and Mental Health reviewed its affiliation with the University of Melbourne. While it remains a completely independent medical research institute with an independent board, in 2012 the Department of Florey Neurosciences was created within the university. This provides selected Florey staff with the rights and privileges of the university, even though they are not staff members of the university. Importantly, the research output of the Florey and other affiliated research institutes contributes to the University of Melbourne's national and international research profile.

Funding for the Florey Institute of Neuroscience and Mental Health is mostly from competitive external grants from the NHMRC and to a lesser extent the Australian Research Council (ARC). Division heads and laboratory heads apply for project funding that is mostly for three to five years. If successful, that funding will cover salaries for themselves and the staff in the division/lab, including early-career postdocs and research assistants (RAs). Funding for their postgraduate research students comes mainly from federally funded Australian Postgraduate Awards (APAs) or Melbourne University Research Scholarships. More senior researchers compete for NHMRC fellowships in order to bring in their own funding and move towards becoming independents researchers, eventually as lab heads. NHMRC Fellowships are funded for five years. Other sources of funding for the Institute are through the state government or the philanthropic sector. The Institute has a fundraising and marketing division that has secured significant funding in recent years.

Some of the issues concerning early-career researchers at the Institute were highlighted in a survey by the Florey Post-doctoral Association (FPA 2011). It found that while most were satisfied with the quality of research and opportunities for internal and external collaboration, most were dissatisfied with communication between the Executive and postdocs. Furthermore, while most were satisfied with opportunities to be first author on papers, opinion was divided about opportunities for last authorship (which is usually the prerogative of the supervisor). Most respondents said that they would like to set up their own lab or become head of a division within the next five years. All were on contracts and were concerned about funding and job stability. They were also concerned about career development, in particular a perceived lack of policy regarding promotion of scientists to lab heads. Twenty-one per cent reported that they were very satisfied with science as a career, and 27 per cent were satisfied, but others were dissatisfied/very dissatisfied (27 per cent) or neutral (27 per cent). The survey found that satisfaction was related to opportunities for external collaborations and career development, ability to influence decisions at the institute and within their group, and recognition of their achievements. Those who were not satisfied with science were more likely to be first rather than last authors on papers. Satisfaction with science tended to reflect perceived recognition and responsibilities, rather than research output. Finally, the areas of greatest perceived need were funding for large, shared equipment; internal promotion opportunities; and longer contracts for postdocs (FPA 2011).

The Florey Institute of Neuroscience and Mental Health has for some time been exploring strategies to improve career progression for its research scientists. It has been particularly interested in exploring initiatives to keep women with family responsibilities in science careers. The issue of effective career progression for research scientists, and especially for women scientists, within the current funding model has been discussed by the institute over a number of years. In 2009, a Women in Science (WIS) group (later renamed the Equality in Science (EqIS) committee) was established to identify issues in the workplace that have impacted on the progression and/or the scientific career development and research outcomes of women researchers at the institute. One of its first initiatives was to write to the NHMRC's Chief Knowledge and Development Officer suggesting policy improvements to NHMRC schemes, which it believed 'would greatly assist in the retention of the best and brightest women researchers'. These suggestions were: that the NHMRC offer paid maternity and paternity leave for all levels of funding; availability of part-time fellowships at all levels to provide better opportunities for women researchers with young children; introducing a new award to accommodate researchers returning from career interruption; allocating a section in funding applications to enable a clear statement of scientific output relative to opportunity for consideration during the review process, because it was not clear how this was currently assessed; and feedback for unsuccessful training fellowship and career development award (CDA) applicants (WIS Group, Howard Florey to Dr Clive Morris, 24 February 2009). The Institute's investment in gender equality is viewed as important and mirrors current thinking in the European Union that this will help to overcome any negative gender stereotypes in the organisation (EU 2012).

Summary

This chapter has identified a significant attrition of women in science research and explored some of the factors that lead women to seek other careers. These factors include: the science funding model with its emphasis on publications in high-impact scientific journals and resulting concept of the successful scientist as a male with no other responsibilities other than doing science; the huge investment required to secure competitive funding and the perceived flaws in the review process; and the lack of acknowledgement that women with care responsibilities may have a different career trajectory, including the preference to work part-time while their children are young, and therefore require different funding models.

The chapter introduced the case study of the Florey Institute of Neuroscience and Mental Health, and outlined its structure and funding. It also reported on the survey conducted by its post-doctoral association and initiatives by the institute to explore strategies to improve career progression for its research scientists, especially for women with family responsibilities.

Chapter 2 will review the literature on building science research careers, especially for women scientists, and chapter 3 will outline the research design. Chapters 4 to 10 will present the results of this research, and chapter 11 is the discussion and conclusion.

CHAPTER 2
Literature Review

The literature on career progression in science, especially for women, is extensive and has received a good deal of attention in recent years from the European Commission which has funded a major study of strategies to change the culture in science research (genSET 2010; genSET 2012).

Gender and Science Careers

It has been argued that gender is a social practice that constructs norms, with class-privileged men as the neutral and objective standard (van den Brink, Benschop & Jansen 2010; West & Zimmerman 1987). Organisations also shape gender identity; as Ely and Padavic (2007, pp. 1138–9) assert, 'organisations serve as historically situated contextual constraints within which people are capable of exercising choice. The interplay between organisations and individuals shapes and reshapes, creates and recreates gender identities in potentially infinite ways'. Taking this perspective of gender as shaped by organisations and social practice, the concept of a successful 'scientist' is portrayed as male, the organisational culture of the science workplace remains male, and assumes, as asserted in chapter 1, that a successful scientist is a monastic male who has no other responsibilities. Etzkowitz et al. (1994, p.65) argued that by accepting 'various parochial ways of conceptualising, investigating, and organising the conduct of science, significant sectors of the population have been excluded from full participation'.

The construction of the linear career path of male scientists being considered the only route to success (Moir 2006) has consequences for women in science. Women often have different career paths due to family responsibilities and may spend several years out of the work force, returning initially to part-time positions. From the construction of the ideal scientist as male flows the notion that work-life balance is only a women's issue in science (Moir 2006) and that the care of children is effectively feminised by organisations (von Alemann & Beaufays 2014). Thus homosociability in the science community—that is leaders supporting and promoting younger employees with similar backgrounds to themselves (Kanter 1977)—favours young male researchers and can have a negative impact on the careers of women scientists (Husu & Koskinen 2010; Murray & Graham 2007; Wenneras & Wold 1997).

Key Issues in Women's Career Progression in Science

This literature review provides a broader context by examining career progression for women academics in science disciplines as well as those who embark on research-only careers either in universities or research institutes. Key issues in career progression for women in science identified in the literature review are examined below.

One key issue is that career progression for women needs to be examined within the broader context of their secondary school experience in STEM subjects, and their later experiences as undergraduates and postgraduates (Bell 2009; Hatchell & Aveling 2008). It also needs to be examined in relation to the relative status of particular fields of science (Yu & Shauman 2003; Ceci & Williams 2011).

A second issue is that the gendering of science careers—having been established during PhD candidature through lack of support and mentoring, particularly in relation to advice about career paths and in the early career phase—persists and is consolidated throughout the careers of women scientists (Bell 2009; Dever et al. 2008; Asmar 1999; Birch 2011; van den Brink 2009; Hatchell & Aveling 2008; Etzkowitz & Kemelger 2001; genSET 2010). Not surprisingly, women are a clear minority in leadership and senior management positions of science institutions (genSET 2010), as well as more generally in the corporate and public sectors in Australia (CEDA 2013).

This different—and gendered—experience of building a science research career needs to be closely examined. Women scientists experience both external and internal barriers in building their careers. A major external barrier is NHMRC funding. It has been demonstrated that decisions about funding health and medical research are somewhat random and that applicants bear most costs because of the length of time needed to prepare applications (Graves et al. 2011). Importantly, once any researcher (male or female) has a career break they fall behind their peers in competing for the same funding, as the main assessment criteria is output; that is, the number of papers and publications in top scientific journals with a high-impact factor (Bell 2009). However, the NHMRC agreed in 2011 to consider any nominated five years of an applicant's career rather than the previous five years, and it also agreed to monitor gender issues in general (*The Conversation* 2011). But one weakness of these NHMRC reforms is that there are no comprehensive guidelines or training provided to reviewers about how to assess output listed in grant applications relative to opportunity. More fundamentally, women with career breaks are competing with colleagues who have had continuous careers and consequently have greater research output. Internal barriers include different treatment, discussed later in the chapter, attitudes and lack of support for female researchers especially during career breaks, lack of funding support if a grant is not renewed, and under-representation of women on committees and at senior executive level (genSET 2010).

A further issue is that financial reward and stability are crucial to achieving a more gender-balanced workplace. There are now fewer men doing PhDs in some key science disciplines (Bell 2009), and many of these are looking for careers outside science research. It has been demonstrated across a number of professions that where prestige and rewards are insufficient, men move to higher-paying jobs and women take their place in the queue (Reskin & Roos 1990; Peterson 2014). In biological sciences, male postdocs often move into positions in industry or overseas that provide more financial reward and greater opportunity structures to commercialise their research (Murray & Graham 2006). Women are also likely to move into industry to achieve greater career stability, but more importantly to have better, family-friendly working hours. However, given the difficulty that science research institutes have in keeping male postdocs, it becomes even more urgent for them to foster the careers of their talented women postdocs. Otherwise, national competitiveness in science will diminish. It can therefore be concluded that if there is to be a more balanced workplace, then financial reward and stability are key issues to be addressed.

The issue of choice for women scientists is discussed extensively in the literature. Cox (2008, p. 38) reported that researchers often considered that they must choose between a research career and a family or a relationship. Moreover, Caprile (2012, p.18) argued that the 'myth' of 'total availability in the scientific lifestyle' leads young women to believe that 'science is incompatible with family life, and to feel that they need to leave science careers if they wish to have a family'. However, men are not asked to make these choices (Moir 2006). Thus, the perception that women scientists need to make a choice between a mobile career and family can be a way of filtering out who will succeed in science (Caprile 2012).

Work versus other responsibilities is another key issue for women in science. Evidence of the impact of relationships and children on the careers of women scientists is not clear, but suggests there is not a strong link between children and lower ambition and research productivity, although women, particularly those with children, are less likely to consider a principal investigator position (*Nature Neuroscience* 2010). Caprile (2012) questioned the impact of marriage and children on scientific productivity (see also McNally 2010) and argued that there were more complex reasons for women's disadvantage. This research concurs with other findings that indicate the perception of senior colleagues and peers about the research focus of women with children appear to have a negative impact in the workplace and can become a considerable obstacle to the academic reputation of these women (Moir 2006; Ceci & Williams 2011; Fox 2005; Fox & Colletralla 2006; Hartley & Dobele 2009; Corley 2005; Mavriplis et al. 2010; Lane 1999). Workplaces therefore need to focus on strategies to enable women to get through the difficult period—often in their thirties—when they are trying to juggle family and career. This includes developing clear paths for re-entry into the workforce and more flexible work styles (Jacobson 2013). Moreover, departments and universities should be encouraged and funded to experiment with alternate life course options (Ceci & Williams 2011).

Even women without interrupted careers do not experience rates of career progression comparable to their male colleagues. Caprile (2012, p. 18) found that there was no clear evidence that women without children have better career prospects 'or that they succeed in catching up with men in their careers'. Part of the reason for their slower career progression may be that they experience different treatment in the workplace which impacts on the salaries they receive relative to male colleagues. A study of broad employment trends found that about half the gender pay gap in Australia is explained by different pay rates in different gendered occupations, but the remaining 50 per cent is attributed to discrimination in the workplace (KPMG 2010). Moreover, a recent study that analysed the publication records of biomedical researchers found that working at a highly ranked university—and being male—are predictors of academic success. Female biologists were less likely to become principal investigators than male biologists with comparable publication records (*Nature News* 2014).

Illustrations of this intentional sex discrimination in science laboratories include: women's ideas, ability and leadership efforts often being either dismissed or countered with hostility; professor's attitudes to female compared to male colleagues (for example inviting male but not female colleagues to social functions); and principal investigators running their labs like a private 'fiefdom' (Stark 2008; see also MIT 1999). Sekreta (2006) argued that the professor's power in science laboratories should be decentralised to prevent them having 'complete and total control' over a student's research, future employment prospects and overall career success.

Gatekeeping is a further key issue for women's career progression in science. It operates in research institutes through lack of support for women from heads of laboratories for funding applications and publications, women being less frequently named as last author on articles, and a culture of inclusion/exclusion often based on gender (Bell 2009; van den Brink 2009; Husu 2004; van den Brink,

Benschop & Jansen 2010; Harman 2010; Acker, Webber, M. & Smythe 2010; Holmquist & Sundin 2010; Griffin 2004; Barinaga 1992; Fouad & Singh 2011; Husu & Koskinen 2010; Etzkowitz & Kemelger 2001; Murray & Graham 2007). Being last author on a scientific article is typically the prerogative of the senior scholar, who is not necessarily responsible for doing most of the research or writing but directs the lab in which the experiment is carried out. Wilson (2012) found in a meta-analysis of publication in journal articles that only about 23 per cent of last authors were female. But in the fields of molecular and cell biology, women represented almost 30 per cent of authorships from 1990 to 2010, but only 16.5 per cent of last authors.

Mobility and international collaboration are critical issues identified in the literature for women's career progression in science. Barjak and Robinson (2008) found that successful life-sciences teams have a strong domestic base but collaborate actively enough outside the country to ensure some external involvement in the team. However, 'Extreme management or policy strategies which result in teams which are all domestic or mostly from non-domestic origins are clearly at a disadvantage compared to those leading to an appropriate mix' (Barjak & Robinson 2008, p. 33). Sabatier et al. (2006, p. 312) examined careers of men and women in a French national life-sciences institute. They found that institutional mobility after recruitment significantly slowed down the promotion process, while international and organisational mobility before recruitment sped up career progression. This underlined that mobility was important, but mostly before recruitment. After recruitment, involvement within the institution was more important.

As there is little reference in the Australian literature to the impact of mobility on women's careers in science, this has been explored in the interviews with research staff at the institute. Ackers (2010) examined the contribution of new forms of mobility and particularly of short stays (whereas two-to-three year post doctoral appointments were once the norm) to the internationalisation process that has become so critical to career progression. This illustrates the specific opportunities that certain forms of short-stay mobility present, both in terms of optimising knowledge-exchange processes and internationalisation, and for 'potentially mobile' women and men with personal and caring obligations.

Ackers (2010) reported that highly mobile researchers in her study considered mobility made family life extremely difficult and added: 'The increasing necessity for dual income families, the difficulties in maintaining two careers and the problems encountered in moving families and partners abroad have emerged as clear inhibiting factors'. Zippel (2012) concurs, arguing that international science is the 'new frontier' for women. Ackers (2010, p. 83) concluded that 'these forms of movement are highly gendered and present unique challenges to people with personal and caring responsibilities'. Mobility is often linked to short-term contracts in science research careers. Short-term contracts can encourage mobility between institutions both nationally and internationally and encourage innovation (HoC 2014).

Another critical issue for women in building-science careers is developing strong networks (Wroblewski 2010; Husu 2004; Leden et al. 2007; Faltholm & Abrahamsson 2010; Benschop 2009; Sagebiel, Hendrix & Schrettenbrunner 2011; Sabatier et al. 2006; Fouad & Singh 2011, Bonetta 2010). Networks are essential for scientists, especially in securing funding and gaining promotion. The control and selection function of scientific networks in these appointment procedures are critical (Wroblewski 2010). However, allocation of funds tends to focus on male values and networks (Husu 2004), produces 'a pervasive culture of negative bias—whether conscious or unconscious—against women, resulting in a lack of professional support and networking' (Leden et al. 2007), and results in less access to important research networks necessary for securing funding (Faltholm & Abrahamsson 2010). Benschop (2009, pp. 222–3) argues that these 'intertwined processes of networking and gendering are micro-political processes: they reproduce and constitute power in action in everyday organisational life'. Thus, powerful homophilious networks, or what is often called homosociability, operate by selecting those 'with familiar qualities and characteristics to one's self' (Grummell et al. 2009, p. 335) and can often exclude women.

Several researchers have focused on the importance of women building strong networks and mentoring in order to ensure career progression (van den Brink 2009; Benschop 2009; Faltholm & Abrahamsson 2010). There is evidence that women who have mentors have greater research productivity and career satisfaction than those without mentors (HoC 2014). However, women are largely excluded from vital resources (Morley 1994) as well as key networks that would enable them to build international research profiles. As Wilson-Kovacs et al. (2006, p. 683) assert: 'Informal networks of influence outside women's reach are at the core of their struggle to be acknowledged and treated on an equal footing'. Some women scientists therefore choose to develop all-women networks as a strategy in furthering career progression (Sagebiel 2013).

An additional issue for women in building careers in science is gender bias: the science excellence system that fails to integrate women into scientific networks, lack of gender balance among gatekeepers and application of different criteria (genSET 2010). Sabatier et al. (2006, p. 312) found in promotions processes at a prestigious research institution that: 'it is mainly though internal networks and mentoring that researchers are chosen to be team leaders or deans'. Moreover, Hartley and Dobele (2009, p. 53) reported that women who worked in research partnerships were significantly more likely to submit funding applications with other researchers than those who tended to work alone, demonstrating the importance of formal research networks in developing research careers. Clearly, effective networking—both internally and externally—is critical for securing funding, developing research collaborations and securing promotion.

Finally, at a national level, a key issue for women's careers in science is improving research quality by addressing gender analysis in scientific research (genSET 2010). The UK House of Commons committee (HoC 2014, p. 10) suggested that national academies, learned societies and research funders review how gender analysis can improve research findings within different STEM disciplines, and that 'research funders should encourage consideration of gender dimensions of research from funding applicants'. Moreover, the evidence base in Australia needs to be improved to provide consistent, systematic reporting of gender data in the sector on the part of the major research and research funding agencies (including The Commonwealth Scientific and Industrial Research Organisation (CSIRO) and the NHMRC, centres of excellence (the Learned, the Cooperative Research Centres, ARC Centres and Networks) and industry. In addition, the Australian Bureau of Statistics and Office for

Women need to generate data sets that link participation to innovation in keeping with international practice (Bell 2009). As well, further research on gender and science is needed along with more specific targeting of gender in research issues (genSET 2011).

Initiatives to Improve Women's Careers in Science

The literature identifies a broad range of initiatives that can effectively improve the careers of women in science. These include:

- scientific institutions setting targets to improve gender balance and action plans to reach these targets as part of their gender strategy (genSET 2010);
- funding opportunities to specifically target gender in research issues; training researchers on the gender dimension; and facilitating cross-sector collaborations (genSET 2011); and integrating gender monitoring into quality-assurance processes of funding bodies (Husu & Koskinen 2010);
- high-quality affordable on-site childcare (*The Conversation* 2011);
- universities and research institutes supporting a part-time tenure track (WiS 2009) and providing flexibility for women with family responsibilities to move from part-time to full-time positions (Ceci, Williams & Barnett 2009). While a part-time contract role in a strong research group might bring less prestige and security, it would allow women to develop their expertise and build a track record in preparation for a successful tenured position when their children are older (O'Brien & Hopgood 2011);
- introducing targeted measures to qualify women for professorships (Lind & Lother 2005);
- fellowships with extra resources for top female scientists (these have been introduced by the ARC for their Laureate Fellowships);
- fellowships to encourage outstanding female scientists to take up leadership positions in medical science and to help women reach senior positions;
- adjusting the length of time to work on grants to accommodate child-rearing, no-cost grant extensions, supplements to hire postdocs to maintain momentum during family leave, grants for re-tooling after leaves of absence, couples-hiring, and childcare to attend professional meetings (Ceci & Williams 2011). CSIRO,

Australia's largest employer of researchers, made a commitment in 2011 to increase incentives to encourage women to return to the workforce after a period of maternity leave (*The Conversation* 2011);
- an allowance for outstanding female postdoc fellows and female laboratory heads to assist with the cost of childcare for pre-school-age children;
- family rooms in which to breast-feed infants and express and store milk, and offices that parents can use for emergency/occasional care of their infants and children;
- meeting and travel support to enable outstanding female postdocs and female laboratory heads to join peer-review committees, speak at scientific conferences, and accept invitations to participate in other academic activities;
- technical support during maternity leave;
- additional time for contract renewal;
- family-friendly meeting times and flexible working hours;
- mentoring; and
- leadership workshops (WEHI 2012).

Summary

This chapter examined key issues in the literature in relation to women building scientific careers. These included: the gendering of scientific careers starting at PhD level; external barriers, particularly the funding model, and internal barriers (lack of support and direct and indirect discrimination); choices or lack of choice; work versus other responsibilities and the impact on careers; mobility and international collaboration; and building strong networks. The literature also identifies initiatives to improve the careers of women in science. These key issues and initiatives in relation to women building scientific careers will be explored further in this book.

The next chapter will outline the methodology used in this study.

CHAPTER 3
Research Design

Background

Careers in science research can often be difficult for women. While women in 2009 comprised 67 per cent of scientific staff at what is now called the Florey Institute of Neuroscience and Mental Health, Parkville, and their representation as PhD students was relatively high, they were particularly underrepresented among the institute's senior faculty, as indicated in Table 3.1 below.

Table 3.1: Staffing Profile at The Florey Neuroscience Institute, Parkville, 2009

Position	Total	Number of Women
Senior faculty	25	2 (8%)
Postdoctoral fellows	59	30 (51%)
PhD students	60	38 (63%)

Source: Florey Institute, Women in Science—Activities in 2009

A survey of career opportunities conducted by the Women in Science group at the Florey Neuroscience Institute in Parkville was completed in 2009 by 39 respondents, 77 per cent of whom were women. The findings included: 30 per cent of respondents had had a significant disruption to their scientific career due to maternity/paternity leave or illness; 71 per cent said this had a significant impact on career progression and negative financial implications; students and early career researchers considered that mentoring programs were inadequate; others highlighted lack of part-time career opportunities and difficulties in achieving independence as senior researchers; and others reported lack of career opportunities/progression because of gender.

In 2010, several members of the EqIS committee (formerly Women in Science) at the institute approached Dr Kate White, an established gender-in-higher-education researcher, to review the results of the above survey. Discussion ensued regarding strategies that could be implemented to achieve organisational change and thus ensure better career opportunities for staff, especially women, at the institute. Following these preliminary discussions in October and November Kate left for the UK where she was finalising a book in collaboration with Professor Barbara Bagilhole on women in management in higher education published in the UK and US in April 2011 as *Gender, Power and Management: A Cross Cultural Analysis of Higher Education* (Palgrave Macmillan). Meanwhile she undertook a comprehensive literature review of the issues in relation to career progression for women in science. Members of the EqIS committee again met with Kate in November 2011 to discuss the literature review and to draft a project proposal.

Kate, as the Principal Investigator (PI), developed the research design in consultation with the Scientific Director and the EqIS committee. An ethics application was approved by the Human Research Ethics committee at the University of Ballarat (now Federation University Australia) on 16 February 2012, and on 22 March the project proposal was endorsed by the Scientific Director and the EqIS committee, and a timeline for the project was developed. The only substantial modification to the timeline was that a proposed seminar did not take place in September 2012 as the PI was at the time living in the UK. It was therefore agreed that the draft report would be circulated to the Director and the EqIS committee for comment in November. This book was developed from that research project.

Research Design

Hypothesis

This book will examine the hypothesis that women in science research institutes experience greater challenges than men in building effective career paths.

Aim

The aim of this project was to focus on the Florey Institute of Neuroscience and Mental Health at the University of Melbourne, Parkville, and the Austin Hospital, Heidelberg, as a case study of how to improve institutional practices that can support career progression, particularly for women in science. It will identify current barriers to career progression, including gender and equity and disparity of outcomes, and recommend strategies that can be implemented to address these barriers.

Objectives

The objectives of the research were firstly to examine the broader organisational culture in which women and men attempt to build research careers and its impact on career progression, especially for those who have family responsibilities or who wish to have children. It then focused on external and internal barriers for those in clinical science compared to basic science and strategies that the institute could implement to address these barriers. It was particularly interested in investigating the impact of the organisational culture on women undertaking science PhDs and as early and mid-career researchers. Next, the project examined the gender inequity of women generally not being as well promoted in the field of science, and family responsibilities significantly hindering the productivity of scientists—more often women—and their career progression through, for example, fellowship schemes. It then explored how the organisational culture could be made more compatible with the needs of women and men in the organisation. Finally it analysed effective initiatives to improve career paths for women in science and optimise career outcomes.

Research Questions
The following research questions were developed and formed the basis of the interview schedule used in this project:
- What internal and external factors impact on career progression and how important is financial reward?
- How do career disruptions impact on career progression?
- What are the challenges for building careers for women and men who have family responsibilities or who wish to have children?
- What are the perceived and actual barriers to promotion, and how can measures of success be redefined?
- What role do different research networks play in women's and men's promotion in science research institutes, and do women have the same opportunity as male colleagues to be introduced to these networks? How important is mobility in building these networks?
- How important is mentoring for career progression?
- What are the characteristics of the organisational culture in science research institutes?
- How does the culture of science research institutes impact on the experience of women scientists?
- What strategies need to be implemented to ensure that women undertaking science PhDs and in the early career research phase have the same mentoring and support as their male colleagues?
- What are effective initiatives to improve the position of women in science?

Method
The project used both quantitative and qualitative methodologies to identify issues in relation to career paths for those in clinical science compared to basic science, where women's and men's scientific career paths differ and why this occurs, and how this situation can be addressed.

The initial phase of the research was to analyse statistics on the gender breakdown of all employees at all levels in the institute over the past five years. It then examined recipients of competitive funding by gender for the period 2012–2013.

In the second phase the PI interviewed a sample of forty female and male staff at the Florey Institute of Neuroscience and Mental

Health Parkville campus, which focuses on basic science, and at the Austin campus, Heidelberg, which combines research and clinical practice. A purposive rather than a random or representative sample was used. For sampling in qualitative research the researcher needs to set boundaries (to define aspects of what can be studied within the limits of time and means, that connect directly to the research questions, and will include examples of what they wish to study), but at the same time create a frame to help 'uncover, confirm, or qualify the basic processes or constructs that undergird' the project (Miles & Huberman 1994, p. 27). For this project a purposive sample was used to identify the categories of staff to be targeted for interview, and to enable female interviewees to be matched with male interviewees at similar career stages across the range. The categories identified in discussion with the institute director and the EqIS committee were: PhD students in the third year (or more than three years) of their candidature, early postdocs, mid-career postdocs, and senior PIs/lab heads/division heads, some of whom are members of the institute's Executive. It was decided not to interview RAs as their career paths are different to those of research scientists with whom they work. Theirs is a support role in the laboratory; some may work as RAs before embarking on a PhD, while others— especially women with family responsibilities—prefer the more defined working hours of the RA role. Then, within each of these categories a stratified systematic random sample was used in order to ensure that the sample was not biased. However, in one sub-category—senior PIs/lab heads, which in 2011 had three females and twenty-one males (see chapter 4)—a higher density of sampling was required for the women in order to understand issues for their career progression and underrepresentation at this level.

The career path for science researchers at the institute commences with a PhD that generally takes three to four years fulltime to complete. This is followed by the early-mid career research (EMCR) phase where post-doctoral research officers (ROs) work in the laboratory of a research or senior research fellow who funds the RO, and gradually transition to senior research officers (SROs). EMCRs are researchers within fifteen years post-completion of their research degree (usually a PhD). In the past thirty years in Australia, the average post-doctoral career phase has extended significantly from one to two years in 1980 to in excess of ten years in 2010. Increasingly, this post-doctoral phase 'has become characterised by insecurity in tenure' (McKeon 2013, p. 135). A key transition is from RO to SRO, at which point the postdoc is trying to gain a level of financial independence by applying for competitive external funding. Moving from SRO to Research Fellow generally occurs when a researcher has established their independence and sets up their own lab, as discussed in the next chapter.

Science Research Career Path
PhD → RO* → SRO* → Research Fellow → Senior Research Fellow
* early-mid career researchers

A list of current employees by level was provided to the PI by the Manager of Human Resources. The PI then chose every fourth name on the list. These people were contacted by email. The email invited them to participate, explained the nature of the research project and advised that a report would be prepared for the institute's EqIS committee. It also advised that the interviews would be tape-recorded and made clear that participation was voluntary, and that refusal to participate required no explanation. Furthermore, it stated that the PI would contact them by telephone to see if they had any questions about the project or their participation in it. Some declined to be interviewed; these were mostly PhD students and postdocs. Few of those at research fellow level declined the invitation. Most asked how long the interview would take and were advised it would be from thirty to forty minutes in duration.

Once a prospective interviewee had agreed to participate in the project, a mutually convenient date was set and a venue arranged. Interviews were mostly conducted in meeting rooms at the Parkville and Heidelberg campuses and the PI arranged the bookings. These rooms ensured privacy during the interview. PIs/lab heads mostly preferred to be interviewed in their office. The PI then sent a follow-up email to the prospective interviewee confirming the date and venue, and attaching three documents: the plain language information statement, the consent form, and the interview schedule (see appendix 1). Division/lab heads received a different interview schedule that asked several questions about the challenges of their current position and about their role in developing career paths for their staff. The follow-up email invited them to contact the PI if they had any queries about the documentation.

The process worked well. Few interviewees changed these arrangements once they had been agreed. Most arrived at the venue ahead of time, several with the interview schedule annotated, saying that they had put some time and thought into preparing for the interview.

Interviews varied in duration from thirty minutes to two hours. Some interviewees had particular issues they wished to discuss in more detail. The interviews of more senior staff were generally longer. These were research scientists who supervised research staff and had wide responsibilities both within the institute and externally as reviewers on panels and collaborators on international research projects. It occurred to the PI that while they spent much of their time mentoring staff, they had few chances to reflect on their own careers, and they therefore welcomed this opportunity.

Few interviewees declined to answer particular questions or requested that the tape recorder was turned off when they wished to discuss sensitive information. The exception was questions about promotion. PhD students and early-career postdocs said that they did not know how the promotion system worked in the institute.

The interviews were then summarised and analysed for dominant themes that included: building science careers, job satisfaction, career progression, mobility in science careers, organisational culture, dual careers, work-life balance and generational issues, different career experiences of those in clinical practice and research compared to those in basic science research, career progression of women and disparity of outcomes, strategies to enable women with care responsibilities to stay in science research, and new models for science research careers.

Selecting quotes from the interviews to be used in the report and subsequently in this book required a good deal of careful thought. There were some excellent quotes that the PI decided after some consideration not to use because they could possibly identify the interviewee. Quotes from interviews are identified by an interview number signifying the order in which they were conducted to ensure anonymity. Once the research report was in draft form the PI contacted interviewees by email to ask permission to use quotes from their interviews. In the body of the email or attached to the email were the quotes and the context in which they were to be used. This process worked well. Many of the interviewees were happy for the quotes to be used without any changes being made. Others sent back tracked changes that often amended the quotes. Some requested that parts of quotes be deleted to ensure that they were not identified. A few corrected their grammar. Several requested reassurance that they would not be identified by name. The PI responded by saying that the only identifier would be the interview number. Only two interviewees requested that all their direct quotes be deleted from the draft report, and one of these agreed that an indirect quote could be used.

Once a book contract had been secured the PI, on advice from the chairperson of Federation University Australia Human Research Ethics committee, emailed all the interviewees to again seek permission to use quotes from their interviews, this time for a book. While the quotes had mostly not changed from those in the original report, often the context in which they were to be used had been modified. All interviewees responded positively and only three interviewees requested amendments to quotes from their interviews.

The third phase was direct observation by the PI at seminars within the institute. This was undertaken with the approval of the director of the institute. The PI was as unobtrusive as possible so as not to bias the observations, and was watching rather than taking part. The focus was on observing certain sampled situations or people rather than trying to become immersed in the entire context. Notes were taken in relation to the dynamics of the observed situation, but not in relation to subjects discussed. The PI had also wished to observe laboratory meetings, and the matter was raised with several lab heads. While there was agreement that this in principle was a good idea, they did not consider that observation was appropriate in their laboratory.

Expected Outcome

The anticipated outcomes of the project were:
- identification of where women's and men's scientific career paths differ, why this occurs, and the tools/strategies that can be used to help address the situation;

- recommendation of strategies for clinical and basic researchers that enhance career progression for women scientists within the institute;
- a final report with recommendations to be presented to the institute's Executive; and
- development of a best-practice model of effective career paths for research institutes.

Significance and Innovation

This research project enabled the institute to examine trend data on the representation of women and men in their staffing profile, as well as the relative success of women and men in gaining funding. The project also examined the impact of organisational culture on career progression for women and men, and recommended strategies for enhancing that culture to ensure the institute's staff have productive and rewarding careers.

This project is significant because it is the first comprehensive analysis of gender and equity, particularly career progression of women and disparity of outcomes, to be conducted in any major research institute in Australia. The findings therefore have relevance to research managers and scientists in a variety of organisational settings, as this is the first study to effectively highlight where women's and men's scientific career paths differ, why this occurs, and the tools/strategies that can be used to help address the gender discrepancies. Having identified the relevant issues, ultimately it was hoped that this study would lead to the implementation of practices within the institute that enhance career progression for women scientists and more broadly to provide a best-practice model for other research institutes.

The EqIS committee at the Florey Institute of Neuroscience and Mental Health were invited to participate in the project's design, and in devising the interview schedule. That committee and the faculty of the institute were then invited to comment on the draft report of the findings, and the recommended implementation phase. In this way the institute had ownership of the key findings and responsibility for implementation of key recommendations.

This project's innovation was in the institute analysing its values and strategic direction. This analysis then became a catalyst for transforming the organisational culture in order to optimise career outcomes for all research staff. The transformative model developed hopefully has relevance to other research institutes both nationally and internationally.

CHAPTER 4
Analysis of the Institute's Workforce

In Australia more women than men are currently enrolled in universities (*Bradley Report* 2009). But their representation as undergraduates is not consistent across all disciplines. They are over-represented in Arts, Education and Nursing, and still under-represented in Engineering and Information and Communications Technology. While women are over-represented in undergraduate studies and even in honours completions, there is a clear cross-over that occurs in some science disciplines at the point of research doctoral completions, with more men than women completing PhDs. Figure 1 below shows academic profiles by gender in the natural and physical sciences in Australian universities. While women have slightly higher representation below lecturer level (level A, in some countries called junior lecturer), from there upwards men are over-represented in the remaining academic levels—lecturer (level B) through to senior lecturer (level C) and above senior lecturer (level D, associate professor, and level E professor). In some European countries level C is called assistant professor, and level D reader.

The flow-on effect of fewer women completing PhDs across all science disciplines is 'the high levels of attrition in the postdoctoral phase of women's scientific careers and the small number of women in senior and leadership roles in the science and technology sector'

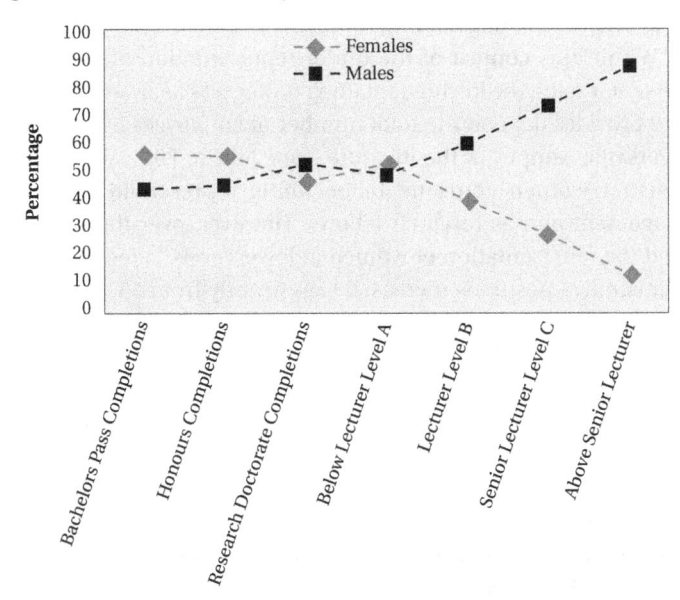

Figure 4.1: Academic Profiles by Gender, Natural and Physical Sciences, 2007

[Line chart showing percentage of Females and Males across academic stages: Bachelors Pass Completions, Honours Completions, Research Doctorate Completions, Below Lecturer Level A, Lecturer Level B, Senior Lecturer Level C, Above Senior Lecturer. Female percentages start around 55% and decline to about 10%; male percentages start around 42% and rise to about 88%, crossing near Research Doctorate / Lecturer Level A.]

Source: Department of Education (formerly DEEWR), Selected Higher Education Student Statistics 2007; DEST Special Report FTE Staff in AOU Groups 2007 (in Bell, 2009, p. 18).

(Bell 2009, p. 10; see also Caprile 2012). However, it should be noted in some disciplines more women than men complete PhDs.

This is not simply an Australian phenomenon. In New Zealand women are particularly under-represented in agriculture, chemistry, computer science, ecology, earth science, engineering, physics, maths and statistics in research careers. And while in general female representation as lecturing staff is increasing, recruitment to full professorships remains low. Moreover, there are very few women as heads of departments in science faculties (AWIS (NZ) 2011). van den Brink (2009, p. 57) has a similar scissor diagram (to Figure 4.1) of the proportion of male and female academics per academic position in the Netherlands between 1999 and 2007. Women have a higher representation than men at Masters level, but from PhD level onwards women's representation declines while men's steadily increases. The little research available seems to indicate that women's experience of undertaking a PhD can be a different one from that of males, particularly in science and engineering (Birch 2011).

Within this context of the under-representation of women in more senior roles, the institute's staffing profile was analysed. Table 4.1 below provides data on the total number of employees by gender at the Parkville campus of the institute from 2007 to 2011, which indicates that women continue to be under-represented in senior management and as research fellows. However, over the five-year period the representation of women at lower levels in technical and administrative positions increased significantly from 55 per cent to 70 per cent. There was a slight increase in the percentage of women as students—from 65 per cent to 69 per cent. This indicates that the institute had difficulty recruiting young men to science research, reflecting a general trend in science research in Australia (Bell 2009). As the McKeon Review (McKeon 2013, p. 134) noted, 'science appears to have diminished in its appeal as a prospective career among young people over the last decade'. It is important to note that this statistical data does not include the workplace profile for institute staff at the Heidelberg campus that combines basic research with clinical practice; this is reported separately.

The representation of women as early-career researchers—these are ROs—has increased from 35 per cent in 2007 to 48 per cent in 2011, although the growth has not been continuous, with women comprising 46 per cent of ROs in 2009 but only 39 per cent in 2010. Research officers are promoted to SROs usually once they become self-sufficient in funding. In some cases a lab head/supervisor may agree to fund them at the higher level, but this is not common. The representation of women as SROs over the five-year period has been uneven: 53 per cent in 2007, 62 per cent in 2009 and 2010, but 57 per cent in 2011. At this level some researchers seek positions in universities rather than aspire to being research fellows because they are looking for job security. Senior research officers are dependent on bringing in their own funding which can be short to medium term. The tenure that a university appointment brings can therefore be attractive. It should be noted that the representation of women as SROs is higher than for ROs. The reason is possibly that some of the female SROs do not have PhDs. While they are integral to the research

Table 4.1: Florey Institute, Parkville, Workplace by Occupational Category by Gender, 2007–2011

Occupational category	Total employees /students 2011			Total employees /students 2010			Total employees /students 2009			Total employees/ students 2008			Total employees/students 2007		
	F	M	%F	F	M	%F	F	M	%F	F	M	%F	F	M	%F
Senior Management	0	6	0	0	8	0	0	8	0	0	6	0	0	6	0
Other Managers	9	7	56	8	7	55	6	7	46	6	7	46	6	0	100
Research Fellows	3	21	14	1	21	5	1	21	5	2	18	10	2	19	10
Senior Research Officers	17	13	57	23	14	62	23	14	62	19	18	51	18	16	53
Research Officers	13	14	48	11	17	39	13	15	46	11	16	41	9	17	35
Students	48	22	69	49	24	67	47	25	65	51	27	65	47	25	65
Tech. & Admin	57	25	70	50	25	67	57	28	66	49	30	62	36	30	55

Analysis of the Institute's Workforce

projects in the laboratory, they are unlikely to secure their own funding and to be promoted to the level of research fellow.

The representation of women as research fellows at Parkville has been uneven, but has been consistently low—from 10 per cent in 2007 and 2008 to 5 per cent in 2009 and 2010, and 14 per cent in 2011. Appointment as a research fellow is very competitive: usually SROs apply externally for these prestigious fellowships. However, it is possible to get promoted as a research fellow internally on the basis of performance through the Florey Scientific Promotions committee. The institute has recognised the sharp decline in representation of women between the SRO and research fellow level and has created the Fred P Archer Fellowship to specifically appoint promising female researchers. In addition, a $5 million endowment is being established for senior female researchers.

The sharp decline in representation of women between SROs and research fellows requires further investigation. While seeking greater job security by taking up a university appointment may account for some of the decline in representation, there are possibly other factors at work.

There is a clear gendering of management positions in the institute. Over the five-year period there were no women in senior management. However, women have been mostly over-represented in other management—managers of finance, HR, fundraising and marketing, and scientific services—although the numbers are small: 100 per cent in 2007, 46 per cent in 2008 and 2009, 55 per cent in 2010 and 56 per cent in 2011.

Table 4.2: Brain Research Institute, Austin Hospital, Workplace by Occupational Category by Gender, 2007–2010

Occupational category	2010		2009		2008		2007	
	M	F	M	F	M	F	M	F
Admin	0	3	0	3	0	4	0	3
Technical & RA	5	7	5	7	5	8	3	5
Postdoctoral	11	3	10	4	12	3	13	1

Tables 4.2 and 4.3 provide statistics on the representation of women at the Brain Research Institute (BRI) and the National Stroke Research Institute (NSRI) at the Austin Hospital, but do not provide a

breakdown of the postdoctoral category into early career researchers, SROs and research fellows, and the definitions used to classify postdocs may differ from those used at the Florey Parkville campus during the same period (email, K Site, 2 December 2012). However, the tables do suggest that at BRI women were under-represented, while at the NSRI women were over-represented in 2009 and 2010.

Table 4.3: National Stroke Research Institute, Austin Hospital, Workplace by Occupational Category by Gender, 2007–2010

Occupational category	2010		2009		2008		2007	
	M	F	M	F	M	F	M	F
Admin	0	7	0	6	1	10	N/A	N/A
Technical & RA	8	29	5	23	4	15	N/A	N/A
Postdoctoral	7	14	7	13	10	9	N/A	N/A
Students (PhD)	4	10	2	9	0	6	N/A	N/A

The breakdown of full-time and part-time employees by gender (see tables 4.4 and 4.5) is interesting. The percentage of both women and men working part-time at the institute increased over the five-year period. The representation of women working part-time at the lower technical and administrative levels was consistently high but declined slightly from 89 per cent of those working part-time in 2007 to 77 per cent in 2011. There were no students undertaking PhDs part-time. The reason is that PhD students are all scholarship holders and for those on Australian Postgraduate Awards (APAs), part-time scholarships are available only for socially compelling reasons (for example, health-related issues, and not if they are working elsewhere). A part-time scholarship holder would receive half the income and they would lose their tax-free status, as they would need to declare the part-time stipend for tax purposes.

Over the five-year period there have been more women than men working part-time as RAs—varying from 100 per cent in 2007 and 2010 to 75 per cent in 2009 and 2011. Few female or male ROs work part-time. In 2007, there were two women and two men working part time, but in 2008 and 2009, there were only men working part-time at that level. And in 2011 there were three women and one man working part-time.

At SRO level women's representation as full-time workers varied from 45 per cent in 2011, to 48 per cent in 2007 and 2008, and 59 per cent in 2009 and 2010. But women had a higher representation as part-time workers, ranging from 67 per cent in 2008 to 88 per cent in 2011.

The pattern changed at the level of research fellows, where, as discussed above, women were significantly under-represented. But interestingly from 2009 to 2011 only males were employed part-time at this level, three in both 2009 and 2011 and two in 2010. And in 2007 and 2008 there were two men and one woman working part-time. This would suggest that the institute supports researchers working part-time at a high level. Some of these males were approaching retirement and elected to work part-time; others held joint appointments between the institute and a university.

Table 4.4: Florey Institute, Parkville, Full-time Employees/Students by Gender, 2007–2011

Occupational Category	2011			2010			2009			2008			2007		
	F	M	%F	F	M	%F	F	M	%F	F	M	%F	F	M	%F
Senior management	0	5	0	0	8	0	0	8	0	0	6	0	0	6	0
Other Managers	8	7	53	7	7	50	5	7	42	5	7	42	5	0	100
Research Fellows	3	16	16	1	18	5	1	18	5	1	16	6	1	17	6
Senior Research Officers	10	12	45	17	12	59	17	12	59	15	15	48	13	14	48
Research Officers	9	10	47	11	15	42	11	13	46	9	13	41	6	15	29
Research Assistant	13	9	59	24	11	69	29	9	76	27	11	71	20	12	63
Students	48	22	69	49	24	67	47	25	65	51	27	65	47	25	65
Tech & Admin	30	19	61	26	18	59	27	15	64	24	14	63	20	15	57
TOTAL	121	100	55	135	113	58	137	107	56	132	110	55	112	104	52

As noted earlier, the representation of women at the other manager level is higher than for senior management. All but one of the women at this level over the five-year period worked full-time and none of the males worked part-time. There were no women at senior management level over the five-year period and only one man worked part-time.

Table 4.5: Florey Institute, Parkville, Part-time Employees/Students by Gender, 2007–2011

Occupational category	2011			2010			2009			2008			2007		
	F	M	%F	F	M	%F	F	M	%F	F	M	%F	F	M	%F
Senior management	0	1	0	0	0	0	0	0	0	0	0	0	0	0	0
Other Managers	1	0	100	1	0	100	1	0	100	1	0	100	1	0	100
Research Fellows	0	3	0	0	3	0	0	3	0	0	3	0	1	2	33
Senior Research Officers	7	1	88	6	2	75	6	2	75	6	2	75	5	2	71
Research Officers	3	1	75	0	0	0	0	1	0	0	1	0	2	2	50
Research Assistants	3	1	75	5	0	100	9	3	75	9	3	75	5	0	100
Students	0	0	0	N/A	N/A	N/A	N/A	N/A	N/A	N/A	N/A	N/A	N/A	N/A	N/A
Tech. & Admin	19	3	77	9	2	82	11	3	79	11	3	79	8	1	89
TOTAL	24	10	71	21	7	75	27	12	69	27	12	69	22	7	76

Casual employment at the institute has been restricted to both lower-level research and to administrative and technical positions (see Table 4.6 below). From 2007 to 2011 there were no casuals above the level of RO, and the numbers at that level and RA level were low. The only significant numbers of casuals were at the technical and administrative level and these have been increasing. In 2007 women comprised only 36 per cent of casuals but by 2011 they comprised 85 per cent.

Table 4.6: Florey Institute, Parkville, Casual Employees* by Gender, 2007–2011

Occupational category	2011			2010			2009			2008			2007		
	F	M	%F	F	M	%F	F	M	%F	F	M	%F	F	M	%F
Senior Management	0	0	0	0	0	0	0	0	0	0	0	0	0	0	0
Other Managers	0	0	0	1	0	100	0	0	0	0	0	0	0	0	0
Research Fellows	0	0	0	0	0	0	0	0	0	0	0	0	0	0	0
Senior Research Officers	0	0	0	0	0	0	0	0	0	0	0	0	0	0	0
Research Officers	1	3	25	0	2	0	2	1	67	2	1	67	1	0	100
Research Assistants	5	0	100	3	0	100	2	1	67	2	1	67	1	3	25
Students	0	0	0	N/A	N/A	N/A	N/A	N/A	N/A	N/A	N/A	N/A	N/A	N/A	N/A
Tech. & Admin	17	3	85	11	3	79	17	10	63	16	14	53	8	14	36
TOTAL	23	6	79	18	7	72	21	12	64	20	16	56	10	7	37

*Casual employment is defined in the institute's enterprise agreement as short-term work of an ad hoc nature, or the replacement of a permanent employee who is on leave for up to four weeks.

Table 4.7 shows the institute's competitively obtained peer-reviewed funding for research program, project, fellowship and scholarship grants, as reported to the state government. Where males and females shared a grant, the amount of the funding was allocated 50–50. The table indicates that far fewer women than men were successful in securing funding. This in part probably relates to the under-representation of women in senior research positions, as indicated in Table 4.1.

Table 4.7: Florey Institute Funding by Gender, 2012–2013

Number of recipients		Total	Amount		Total
M	F		M	F	
52	18	70	$12 744 887	$2 934 253	$15 679 140

These findings are consistent with the success rates by gender in 2011 for all NHMRC grants where there were almost twice as many men (60 per cent) as women (38 per cent) listed as CIA on active NHMRC grants (Evans-Galea 2012). The success rates for NHMRC grants in 2003, 2006 and 2009 are detailed in Table 4.8 below.

Table 4.8: Success Rate by Gender for NHMRC Grants, 2003, 2006, 2009

Grant Type	2003			2006			2009		
	F	M	Overall	F	M	Overall	F	M	Overall
Project grant	20.5%	22.8%	22.2%	17.4%	23.9%	21.7%	22.4%	29.4%	27.0%
Programs	0.0%	44.0%	40.7%	33.3%	47.8%	46.2%	14.3%	21.6%	20.7%
Research fellowship	42.3%	24.0%	31.5%	30.0%	45.9%	41.4%	41.0%	43.3%	42.6%
Career development award	31.1%	20.8%	26.0%	16.0%	34.3%	26.2%	12.4%	16.8%	14.5%
Training fellowship (Australia)	46.9%	31.0%	40.9%	22.4%	14.0%	18.4%	28.7%	32.5%	30.2%
Training fellowship (overseas)	41.3%	46.3%	43.7%	19.7%	32.6%	25.2%	30.6%	29.4%	30.0%
Scholarships	52.5%	56.0%	53.8%	44.0%	55.2%	48.0%	40.1%	47.3%	43.5%

Note: This table shows the success rate of grant applications based on the gender of the Chief Investigator A only. As both Project and Program grants have multiple chief investigators, these data should be interpreted with caution. Those schemes that have only a single named investigator (Research Fellows, CDAs, Training Fellowships and Scholarships) are the true reflection of the success rate by gender.

Source: NHMRC Research Funding (2010). *Facts Book*, NHMRC Publications, Reference NH0138.

Summary

Women at the institute are over-represented as PhD students, with presumably more women than men completing PhDs. This differs from Figure 4.1, which indicated that in 2007 slightly more women completed PhDs in the natural and physical sciences nationally. The representation of women at the early career stage (as ROs) has increased over the last five years, and there are now almost equal proportions of men and women at this level. At the next level—SROs—women have been consistently over-represented during this period. It was noted above that some of these women do not have PhDs, which may be a factor in their remaining as SROs rather than progressing to the next level. However, women are significantly underrepresented as research fellows and senior managers. At the Austin campus the representation of women at NSRI at postdoctoral level is higher than for men, although there is no breakdown available of RO, SRO and research fellow levels. There has been an increase in both women and men working part-time at the institute over the five-year period, suggesting some flexibility in work patterns at some levels. Casual employment at the institute has been restricted to both the lower level research and administrative and technical positions.

In relation to funding, significantly fewer women than men are recipients of competitive funding and consequently women bring in less money to the institute.

Over the five-year period under study there were no women in senior scientific management. However, women were mostly over-represented in lower management levels such as in areas supporting science research. The challenge for the institute then is to provide opportunities for young women to build effective career paths and to provide the support required to enable them to reach senior research levels. Such support includes the institute appointing women to chair internal committees, appointing new female laboratory heads, ensuring greater transparency in the institute's promotion policy, and engaging in discussion about the organisational culture and its impact on the representation of women at senior levels. These issues will be discussed in detail in chapter 8 and recommendations made. It is hoped that such initiatives will lead to an increase in the representation of women as research fellows at the Parkville campus and to an increase in the representation of women in scientific management.

CHAPTER 5
Job Satisfaction

Why do people decide to embark on a career in science research? Is it a passion for discovery or new knowledge, or is it more focused on specific goals and an ability to shape scientific agendas? Is it about financial reward? And what about the opportunity perhaps in the future to become an independent researcher?

Doing the Science

This study was interested in research scientists engaging in discussion about what gave them most satisfaction in their job. All were keen to explain why they were scientists. Their responses indicated an extraordinary commitment to what they did, as well as to the wider scientific field. The gender of respondents in this and the following chapters is mostly not identified in order to ensure unanimity. Moreover, in this chapter and chapter 6 the focus is on perceptions of science research careers rather than any detailed gender analysis of careers. More extensive gender analysis is undertaken from chapter 7 onwards.

Most respondents described the science as providing most satisfaction, as one explained: 'The discovery aspects and finding something that has never been done before, that generally makes your day-to-day work invigorating' (Interview 3).

The passion for what they do as science researchers was palpable. Science was so exciting that several described how much they really liked their work environment and woke up each day eager to go to work, not necessarily for the salary and even though for one there was the logistical challenge of juggling children and work:

> This is definitely what I want to do. I love what I do and I think I am privileged to have a job that I enjoy getting out of bed for every day. I am on track to doing what I want to do … I don't think money is what holds us in the game in science, it's because we have a passion for what we do. In saying that, getting an increment in the salary is always a bonus (Interview 21).
>
> I love answering research questions; I have got a fantastic team that I am working with now, but that is another thing … There are some really big questions we need to answer. I am lucky because I am a clinician; I can actually see a patient and think 'how can I answer those questions they are asking?' I feel I am lucky to be in the position where I can cross that and answer those questions. I wake up in the morning thinking 'how I can get my kids to school so I can get to work?' But looking forward to going to work, very much so (Interview 35).

Part of what respondents described as the science was problem solving, 'I am a problem solver by heart. It is the actual nitty gritty of research and getting to the science behind it' (Interview 9) and 'my area of research is like the ultimate puzzle. I love puzzles … Neuroscience is *the* puzzle to work on. I find this area of research fascinating' (Interview 20; also Interviews 1, 12, 26 and 28). There was enormous creativity involved in the process of problem solving; for example, a respondent described how: 'It is definitely being able to think outside the square, being able to apply ideas to an experimental paradigm, and then test the experiments and then when your ideas work or you develop a new method, it is just a fantastic feeling' (Interview 14).

For some respondents, doing science almost bordered on obsession; for example: 'every now and then you get an unexpected result, which is quite exciting. I kind of liken it to playing the pokies, not that I am a gambler … it is something weird that keeps bringing you back, that makes it interesting' (Interview 29).

Doing science was so important to one respondent that they were prepared to move countries in order to pursue that passion: 'And in fact if it ever got to the point in Australia that I couldn't do my science full time in the way I wanted to do it, with adequate funding, I would move back overseas' (Interview 27).

Several described the sheer excitement of a scientific breakthrough as providing the greatest job satisfaction, as the following respondent explains:

> What I really love about research is the fact that I am doing things that no one has done before, by definition you are investigating new things. So what gives you most satisfaction is that moment when you're looking at data for the first time and getting results … that is the exciting moment (Interview 10; also Interviews 16, 18 and 19).

Other respondents took the view that working in neuroscience at the present time was like being in the right place at the right time; 'the field of neuroscience is expanding and is a fascinating field' (Interview 12), and that now was a 'golden era of neuroscience'; 'This is the great frontier. This will be remembered as the golden era of neuroscience and genetics and a number of other areas of medical research. It is an incredibly exciting time' (Interview 27).

Some respondents were more outcome-focused, citing results and publications as aspects of the job that provided most satisfaction, as one explained:

> Results give me most satisfaction. To be a scientist you really have to have that curiosity to keep you going. You really want to know the answer to the question you are trying to ask of nature. If the answer is positive, when you finally get to the end of the line that creates a story that can get into a journal, it is a great sense of achievement. It's fun (Interview 13).

For another the excitement of results and interaction with their team was clear:

> The science gives me the greatest satisfaction ... I love sitting on the microscope and seeing the data come through for the first time. That is why I do the job. We could all go and make more money elsewhere. The satisfaction of pitching a question, seeing the results come through. I love interacting with my staff and the young people and the motivation that comes from that ... I have a lot of good friends here. It is a nice place to come to every day (Interview 21).

This excitement, then, made the institute a good place at which to work.

For yet others it was the team work and collaboration that provided most satisfaction on a day-to-day basis. One interviewee talked about an excellent research team (Interview 35). Others commented: 'Just working with other people. Doing science you are discovering things. I like the collaborations and working with people' (Interview 11); and 'I still enjoy the international collaborations that I have built up. We have visiting scholars coming from overseas fairly regularly. That is fun and good culturally' (Interview 31). For this respondent, and Interviewee 13 above, doing science was fun, suggesting a mix of excitement and adrenalin rush.

Several respondents considered that communicating the results of scientific research was just as important as the experiments, and more women than men were focused on communicating science: 'I get most satisfaction from talking to people about my science, not actually doing it. It is a shame that communicating science is not rewarded by our system. It is such an important part of being a scientist, and part of our obligations to the community' (Interview 5; also Interview 7); for another it was 'attending conferences and giving talks' (Interview 8). The stronger emphasis of some women on communicating science was consistent with the findings of Palacin et al. (2013) that women scientists were more interested in social innovation than developing technical solutions.

The views of these and other respondents were a far cry from Charlesworth et al.'s (1989, pp. 272–3) observation of another Melbourne research institute about 'the contradiction between the ideal norms of science (disinterestedness and public access) and the reality of science dominated by a competitive and individualistic ethos'.

Support and receiving mentoring from managers and colleagues, as well as providing mentoring, was also highlighted by several respondents as an important component of job satisfaction: 'I gain a great deal of satisfaction from interacting with my peers; sitting around a table and troubleshooting about where we are going next, sharing excitement about a new result and also sharing the disappointment when a result isn't as we expected' (Interview 5).

> As a group I find it very collegiate. It is not people defining little islands for themselves, it is very interactive. It exploits the fact that there is so much expertise in having that work well, you have to have people with different skill sets. It can only work as a group exercise (Interview 17).
>
> I like to change things and build groups. I find that really satisfying to do that. I like challenging old ideas and norms ... A part of building the group is mentoring. I have always loved mentoring (Interview 23).

What emerges here is a picture of an organisational culture which works in a collegial and collaborative manner and seeks to maximise the expertise of its research staff.

Training the next generation of science research leaders was also important for senior scientists: 'Student training gives me most satisfaction. Interfacing with all these people who have published PhDs ... for me that has been the most rewarding thing of my whole career' (Interview 37); and 'I like working in a team environment, training new students and being part of a team solving problems. I like that as well' (Interview 26). Another elaborated:

> I enjoy interacting with the student body and doing various admin tasks on their behalf ... We have been very successful in attracting new students. And we have now got

some postgraduate students who have stayed on as post-doctoral fellows and are doing further research. I enjoy interacting with my staff and my students ... We have international students in the team; and that is good for our local students. And I enjoy the cultural aspects of all that as well (Interview 31).

Again, this narrative is full of energy and optimism. It mentions once more the fun involved in good science and fostering international networks.

It is interesting to contrast the views of these respondents with the findings of *Nature*'s (2010) first ever salary and career survey, looking at overall career satisfaction and the factors that contribute to it by career stage, gender and geographical region. More than 10,500 scientists from dozens of countries, including Australia, responded. The survey found that 'guidance received from superiors or co-workers' was the biggest influence overall on satisfaction levels, and suggested one possible conclusion was 'that scientists crave guidance and mentoring, seeking assurance from others that they are likely to learn and progress—and they may place a higher premium on mentoring as science careers become increasingly competitive'. Certainly in this study there was much discussion of working in teams led by lab or division heads, and having a rich resource in the colleagues within and beyond the team. A study of scientists working in industry also rated highly the feeling of 'being part of—and contributing to—a team' (*Research & Innovation* 2005). The importance of being mentored and mentoring others in building scientific careers will be analysed at length in chapter 7.

Financial Reward

The *Nature* (2010) survey found that salary was the second-biggest driver of satisfaction for scientists. This was markedly in contrast with the findings of the present study of the institute where respondents were asked: 'How important is financial reward?' Most considered that it was relatively unimportant, as science by its nature was not a high paying career, as one respondent put it: 'If financial reward was important we wouldn't be scientists' (Interview 22).

While those interviewed in this study required sufficient remuneration to pay the mortgage or—for younger scientists—the rent, it was not what motivated them to come to work each day; rather it was the science: 'Money was never that important to me at all because I really liked the job' (Interview 37). 'I don't see financial reward as the be all and end all. In life you have to do things you are passionate about, and that is my primary aim. I get paid enough' (Interview 14; also Interviews 10, 11 and 29). Another valued the strong workplace culture: 'I value workplace enjoyment over financial reward' (Interview 26). Others also did not regard financial reward as 'a motivating issue' (Interview 28; also Interview 31).

Several respondents were acutely aware of the relatively poor salaries in science compared to those of friends who went into highly remunerated legal and financial services roles. One commented 'I don't think it is helpful to compare science salaries with city banking salaries' (Interview 9) and another asserted:

> To be frank the financial reward is not very high compared to all my other friends who did not stay in science. Then they see the competition I am in and they think I am stupid. So the financial reward is a bonus if you get paid more than what you need. In Australia that is a lot of money, especially if you want to buy a house (Interview 20).

One other reason why financial reward was not important for several respondents was that they were in dual careers and their partners earned as much as or more than they did. The challenges of dual careers for younger scientists will be discussed further in chapter 10. However one respondent did value financial reward: 'I think decent financial reward is important. So part of the reason I wanted to be promoted was because you get paid better' (Interview 40).

In summary then, contrary to the findings of the *Nature* survey, financial reward was relatively unimportant to most respondents in this study. They accepted that science was not highly remunerated. Nevertheless this did not impact on their passion for and commitment to science; money was a means to an end—providing the means to pay the rent or buy a house—but certainly not an end in itself. This finding also resonates with Blackmore and Kandido's (2011, p. 408) argument that some academic motivations 'are about intellectual positioning rather than financial gain'.

Becoming an Independent Researcher

In this *Nature* survey 'degree of independence' was ranked third as a determinant of overall satisfaction. Independence was certainly important to those in the Florey Institute of Neuroscience and Mental Health study. Early career postdocs were asked if they wanted to become independent researchers; that is, if they wanted to submit applications to secure their own funding, rather than depend on funding secured by the lab head, and to then build their own research team which initially might comprise only of an RA and an honours and/or a PhD student. Becoming an independent researcher is a fiercely competitive and political process that relies heavily on securing external funding, as discussed in chapter 6, and can unleash underlying power struggles. Charlesworth et al. (1989, p. 270) observed in another research institute that power plays a large part in the scientific process through 'staking out and legitimising a scientific field'; setting a scientific agenda within a given field, 'securing government and private patronage'; and 'ensuring funding, promoting personal and institutional glory, and winning scientific "races"'. Nevertheless, there is often room for more junior researchers to demonstrate independence by establishing their own labs. Unless they move, a bottleneck can develop in physical space requirements.

Not surprisingly then, one postdoc saw this process of becoming an independent researcher as being quite daunting, noting that it was a really competitive environment and no-one was holding their hand, and describing their team as 'very independent'.

> Yes, I would like to be an independent researcher. The barriers are I don't have the CV to become one. A lot of the work I do, there are other senior researchers who are more likely to get funding, which is not a bad thing. I kind of feel in my team there is not a great deal of support. You have to find the grants and apply for them. And if you miss them, that's your problem. We are very independent. No one comes and taps you on the shoulder and says 'there is this

early postdoc and you should look at it'. It is fair enough as long as you realise you should be looking out for opportunities (Interview 9).

The challenge here was to fully understand the process of looking for opportunities and then taking initiative. However, another post doc who was further advanced in their career viewed the goal of becoming an independent researcher with less trepidation than the respondent above, and seemed to be already implementing strategies to achieve such independence:

> I would say I am mid-career. I do want to become an independent researcher. I have got things up my sleeve I want to look at. And it is starting to happen. I have got a few RAs and students that are working on my projects. I don't feel like there are massive barriers that are imposed from above. If you have got interesting work you are more likely to be successful in applying for grant funding. And when you get money you can employ people on your projects. And the institute can help or hinder that. We got funding last year because we had a good idea and we had access to an excellent facility on site (Interview 10).

There were a few respondents who had been recruited to the institute during a phase when it embarked on a strategy of picking winners, described in chapter 6. One of these had confidence that they had the resources and support from management to become an independent researcher:

> I definitely want to become an independent researcher. I think my current position is the opportunity. I have been given all the resources to make that happen. And the barriers are the matter of publishing papers. My salary is covered by a fellowship with NHMRC and my research budget was part of the package given to me to come back, and it is on the proviso that I apply for grants. They are underwriting me until I can get a grant (Interview 14).

In summary, some early career researchers wondered if they might ever reach the point of becoming independent, while those who had a few more years' experience were already developing strategies to achieve this goal and were confident of doing so.

Flexibility and Intellectual Freedom

One of the strong themes as a determinant of overall job satisfaction in the current study—but not in the *Nature* (2010) study—was the sense of flexibility. Indeed many of those interviewed valued this flexibility over financial reward, as the following respondent described:

> The other thing I like about my job is that it is flexible. I can come in when I want. It makes you feel very in control of your work. It is a nice feeling. Leading up to conference deadlines we all work long hours … I like the flexibility. Even just to be able to pop down the road to have a coffee with colleagues (Interview 29).

The link between flexibility and being in control of the research was an interesting comment here.

Flexibility extended to working hours, and could be rather unorthodox, depending on the particular lab, as the following respondent indicated: 'Yes, there are flexible work arrangements. We are very, very flexible. Everyone is very flexible. We have [some] students that come in at 7 pm and work until 2 am. So we don't have "the lab opens at 9 am and closes at 5 pm" mentality' (Interview 2).

But mostly flexibility was around being able to work at home on occasion if a parent needed to be with a young child (Interview 39), or when preparing grant applications (Interview 33). One interviewee who had previously worked in industry noted that: 'Science can be flexible, allowing people involved in academic research to make it fit their day' (Interview 24).

Hand in hand with the notion of flexibility was that of the intellectual freedom of science careers. As one respondent put it, 'I would love to remain in a career that gives me this much freedom. In order to remain free and to be able to choose the projects I want to work on I need to be competitive' (Interview 5). Another saw this freedom as having a job that interested them 'and gives me some independence of thought that is pretty rare in a job. So I prize having that over having more money' (Interview 10). These views are consistent with González Ramos and Vergés' (2012) research which found that scientists particularly valued autonomy and flexibility of working time. A third respondent compared this intellectual freedom to being an artist:

> When I talk to lay people I say [doing science] is a bit like being an artist, they can go out and paint whatever they like. A scientist can go out and research whatever you want, obviously it needs to be interesting to other people so that they will fund it; you are constrained a little bit. Within that constraint there is a lot of intellectual freedom (Interview 15).

Thus, the flexibility of working conditions and intellectual independence were highly prized by respondents in this study, and more than compensated for salaries that did not match those of friends and family working in the corporate sector.

Summary

This chapter has demonstrated that job satisfaction for research scientists is about doing the science. They are passionate about what they do and see it as exciting and fun. Financial reward was mostly considered not important; if it was they would not be in science. Achieving a degree of independence was rated highly by most young researchers. Moreover, flexibility in the job and intellectual freedom were strong attractions. The relatively high job satisfaction was therefore considered to be a defining feature of research science.

The next chapter will further analyse how research scientists build their career paths.

CHAPTER 6
Building Career Paths in Science Research

Introduction

Ten years ago *Naturejobs* began publishing accounts of the hopes, frustrations, scientific victories and career defeats of graduate students. Of the twenty-four graduates and postdocs who kept journals between 2004 and 2009, the status of twenty-one of these was reported in 2011. Twelve—of whom five were still postdocs—were working as researchers at academic institutions or non-profit organisations. Of the other nine, three went into industry, one into government, three were engaged in science writing or communication and two had left science altogether. But 'all shared a willingness to accept personal sacrifice and an ability to adapt to changing circumstances' (Smaglik 2011).

The *Naturejobs*' findings demonstrate that careers in science research are notoriously difficult and that only some of those completing PhDs in science will become research scientists. So why is there a high attrition rate of PhD graduates from science research careers? Factors include, as described in chapter 1, the funding model for science research in Australia which creates a highly competitive work environment in which lab and division heads are largely responsible for securing salaries for their research scientists; a long hours work culture required to be competitive; support and mentoring or lack of it; and the working environment. But universities must also accept responsibility for the high attrition rates. The House of Commons report (HoC 2014, p. 48) asserted that some universities 'appear to be too content to devolve responsibility for working hours, careers support and promotion down to research groups'.

In science research institutes such as the Florey Institute of Neuroscience and Mental Health, senior scientists are responsible for bringing in their own salaries and those of their staff through winning competitive funding. Charlesworth et al. (1989, p. 273) describe the 'game' in science research institutes as competitive collaboration with reputations to be 'won or lost' and add that the game: 'is essentially a struggle for authority. If you can so define the field that your own skills, techniques, instrumentation and theory are the ones that produce the results that others want, then you can ensure greater and more reliable access to funding, research students and publication'. The role of funding bodies, therefore, is critical for science researchers building their careers.

Charlesworth et al. (pp. 272–3) also talk about 'the contradiction between the ideal norms of science (disinterestedness and public access) and the reality of science dominated by a competitive and individualistic ethos' and the lack of 'any real debate' about the very important issues that commercialisation of biotechnology, for example, raises. There was some agreement among respondents in this study about the fiercely competitive nature of science research. As one described it, 'there is no doubt that this sector is Darwinian in nature. It is survival of the fittest ... The institute as a not-for-profit has to make some difficult decisions about its own staffing profile and how it retains and attracts the best' (Interview 36).

In this context, the PI talked with respondents at the beginning of the interview about their career paths in science. They were asked to describe their present role; what factors or people were most supportive in getting them into their current position; what factors had been less supportive; if the current position was their preferred career path in science; and if not, what internal and external factors had impacted on their career.

The responses varied according to the particular stage of a respondent's science research career. It is noted that careers do not develop in isolation. 'Rather they are a function of the context in which the individual is located' (Riordan 2011, p. 111). This chapter will therefore analyse respondents' views on career progression under the following headings that delineate the various career stages of a research scientist: PhD students, early career researchers, SROs and research fellows (which include lab and division heads).

PhD Students

PhD students in the third year of their candidature and those beyond the third year were selected for interview. The standard duration for PhD candidature is three years, but some students may require an additional six to 12 months to write up their thesis. APA scholarships fund PhD students for three years full-time, at the end of which they can apply for a six-month extension of the scholarship. So the students interviewed were in the later stages of their PhD and presumably thinking of the next career steps. Most indicated that family, friends, colleagues with whom they did their undergraduate degrees and, more recently, their supervisors were supportive influences. At this stage not all were strategic about their career planning. One described how 'I sort of fell into it. I really enjoyed my Honours and decided to continue on with it. Whether I started out planning I was going to be a research scientist, I don't think so. But it is definitely what I enjoy to do' (Interview 1).

Some were attracted to the institute by its reputation, including an international reputation: 'I know the reputation of the institute is fairly well renowned. So for me it looks really good that I have done my PhD here; I would say that is a good start for my career' (Interview 1), and 'the institute is very good for its students' (Interview 6).

Several were clear that they wished to embark on a research career; for example, 'I am very keen to continue in the research path. At the moment I plan to do a postdoc in the research area' (Interview 12). But others were reassessing if a career in science research was for them: 'At the moment I don't know if it is the preferred career path. I did want to stay in science for the rest of my life. But being in science and seeing what little job security there is and how little postdocs get paid, I don't know' (Interview 8).

It is worth analysing in some detail the reflections of one PhD student who was at a stage of strategically planning the next career steps, once the PhD was completed:

> Continuing on this research path in science isn't really made very easy for anyone in terms of support and getting positions. So I don't know how feasible it is going to be … The more I go through, the more I realise just how difficult it is going to be in this career.
>
> I think we are pretty lucky here compared to other institutes. There is a pretty good community on the student level and more generally. They really do try and look after their own scientists as well.
>
> Going into my PhD I was not informed about what the career path is post-PhD which I am learning about more and more as I am going through. So while I thought I would continue in research, it has become more and more apparent it is not always that easy. So I am starting to keep my eyes open for different opportunities even though I have not given up on research yet. Of late my supervisor and I have had more mentoring meetings and discussing what I might do in the future and discussing possibilities. I have been looking at different science related careers, maybe in industry or something else (Interview 38).

The key points of this narrative were growing awareness that a science research career would be difficult, the presence of a really supportive research culture in the institute and an appreciation of strong mentoring from their supervisor about career options. Clearly this PhD student and others interviewed were provided with the requisite guidance to enable them to make informed decisions about their future career paths.

Early Career Postdocs

Early career postdocs are generally research scientists who have completed their PhD within the last five years (Bazeley et al. 1996). This differs from the definition provided by the McKeon Review (2013) of early-mid career researchers, which includes ROs and also those who have transitioned to SROs. In the early career phase workers are in a process of adjusting to full-time employment, accepting responsibility, achieving acceptance at work, 'developing special skills, balancing individual needs with organisational demands and deciding whether or not to stay in the organisation' (Riordan 2011, p.111). Women in the early career phase are often 'establishing career goals and reputation, developing structures for childcare and managing their households, and establishing a life structure to facilitate the resolution of life and career issues' (Riordan 2011, p.111). Moreover, the current generation of early career women are more likely to earn comparable salaries to those of their partners/spouses than earlier generations of women in this career period (Gordon & Whelan-Berry (2004) quoted in Riordan 2011, p.111).

In the Florey Institute of Neuroscience and Mental Health the average age of early career postdocs tends to be twenty-five to twenty-seven, although if women have had career interruptions it may be older. They are funded by their team or laboratory leader with the expectation that within three to five years they will secure their own funding. Insecurity of funding is one of the greatest difficulties that early career postdocs experience. The McKeon Review (2013, pp. 134, 136) asserted that they needed 'increased certainty of career progression, appropriate training and mentoring for those with talent and enthusiasm' and, as named staff budgeted for on a NHMRC grant, they needed increased salary levels 'to attract and retain good researchers'.

In contrast to some of the PhD candidates interviewed in this study, early career postdocs appeared to be much more strategic in developing their career paths. They had already made a decision to embark on a research career and had been supported by the institute in making that decision. As one explained:

> This is my preferred career path. I moved purposely into this field. The PhD was what I was interested in doing in exposure to neuroscience and continuing on into my postdoc position was exactly the direction I was wanting to go, combining my previous skills with my interest. I was quite fortunate in that career progression (Interview 3).

Other respondents also spoke about a supportive environment which included good supervision, mentoring and peer support:

> A lot of people find they are really burnt out after their PhD. I guess it is my boss who has been a good supporter ... In addition I have a lot of people who went through with me. Some have gone overseas. But it is important to talk with them (Interview 5).

A second respondent commented: 'My immediate supervisor and mentor was very supportive, as well as my previous group leader. What we do is fairly specialised, so it is advantageous for the group to hold on to people, I suppose' (Interview 22).

Another early career researcher talked about supportive parents and good supervisors in their Honours year and a PhD that led to high research outputs:

> I did get a scholarship that enabled me to study in a lab and then I went on to do my Honours in that lab. I got another vacation scholarship that led to my starting my PhD. I was quite successful in my PhD and I managed to publish three papers, which is quite exciting. There are many of my colleagues who have got fed up with it and moved on to more commercial work (Interview 29).

This narrative emphasises support, funding (scholarships) and success in publishing as key transitions in their career path in science. But the respondent is aware that not all contemporaries in the institute have had such a good run and that some have become dispirited and left research.

The more strategic young researchers were eagerly anticipating the next career move and were anxious to clear the next hurdle. For example, one explained:

> I would have liked to have been successful in my application for an early career fellowship. Being a postdoc in a research institute such as this one is definitely my career

path. I could be slightly one leg up if I had received this fellowship. Some of the factors [are] internal, I could have worked harder. As far as external reasons why—bad luck with experiments ... No individuals or processes have stood in my way (Interview 26).

But not all early career researchers were strategic and pleased with their career progression to date. One had no clear direction: 'I am one of those people who floats without much of a direction and if I am happy in the position I will stay there' (Interview 9). Another appeared to be drifting and was not happy with their career:

> I don't have another career path that I am not taking. Whether I am satisfied with my current career path? That is another question. It might be related to dissatisfaction with my career in general. I don't think there has been anything stopping me taking the career path of my choice. The top priority is having a job (Interview 11).

To summarise, mostly the early career researchers interviewed enjoyed their jobs and were happy with their career progression to this point. They considered they were supported by their supervisors and fellow researchers in the laboratory, and anticipated staying in this field. The more strategic were already looking towards the next career move. But at least two respondents were not strategically planning their career path in science research. It would be interesting to speculate on whether or not they would remain at the institute, or would instead move to industry or academia. The UK Science Minister, the Rt Hon David Willetts, recently described the life of post-doctoral researchers as 'pretty tough' and supported the notion of the PI/lab head having an obligation to think about the postdocs' long-term interests, and advising them what to do next (HoC 2014).

Senior Research Officers

Senior research officers are usually three to five years beyond their PhD, and in order to be promoted to the top two levels—SRO 1 or SRO 2—the institute's enterprise bargaining agreement (EBA) stipulates that they need to secure their own funding. At this level they are looking at moving towards becoming independent researchers or are already independent, having secured funding that will pay their own salaries and perhaps those of an RA. They would also be co-supervising post graduate students, with a more senior researcher being the principal supervisor. The insecurity of funding can impact on the type of research pursued. Mid-career researchers tend to favour 'conservative, short-term projects' rather than 'research which may have a high level of risk, but might also carry a greater chance of producing innovative outcomes' (McKeon 2013, p. 136).

Most of those interviewed considered that they had been well-supported in building their careers. For example the following two respondents talked about support from their supervisors to work at the institute, after having postdocs positions overseas:

> My current position is more than my preferred career path in science. I thought I would be coming back from overseas and working for someone else. It is a bonus that I have been given a bit of independence. The reason my current position came about was [due to] the contacts, my old supervisors, and people in the institute. They were very supportive and wanted to have me come back (Interview 14).
>
> [Support from] my principal supervisor. I got a fellowship to return to Australia to work with him. And he has been very supportive along the way. Otherwise I wouldn't say I have had a huge amount of support from the Florey. During my time at the Florey the postdocs themselves have instituted a lot of things that have resulted in more support for the postdocs from the Executive. My direct mentoring is from my principal supervisor (Interview 19).

Working overseas in the early stages of a post doc is generally considered an essential component of building a science research career, as discussed in chapter 7. For Interviewee 14 this experience had put them in a strong position and led to their being given a degree of independence, but Interviewee 19 had not felt hugely supported in the institute except for the supervisor's support and talked about their career path as being slow.

Unwavering support from the supervisor, together with serendipity, were strong themes of the following narrative. This respondent had the opportunity to step up, and the supervisor ensured that they had sufficient support and mentoring to succeed in the new role:

> I started at the Florey six years ago straight after finishing my PhD ... My role has now changed a lot since then; I was RA then RO, very junior. Then my boss left ... and I have taken on more of a managerial role in my lab ... my supervisor and mentor ... made sure the environment was the most supportive it could be, and that I wasn't left floundering in the Florey. He was probably the most influential and supportive person getting me into this role at the moment, and making sure that I am represented properly in the consortium, and also in the Florey making sure my career progresses (Interview 39).

Nevertheless, doing science at this level was extremely demanding, even when receiving support: 'There is always pretty tough competition ... I have always felt pretty well supported. It is constructive criticism to say "this is where you went wrong"' (Interview 20).

Thus, these SROs were increasingly strategic in building their career paths. Some had been recruited to the institute while they were working as postdocs overseas, a strategy that some interviewees described as picking winners. Another respondent had strong support from their supervisor to step up to a more senior role. It would seem unlikely that scientists at this level would leave research careers.

Research Fellows

At fellowship level in the institute all fellows have secured independent external funding and are mostly running their own laboratories. Once they have fellowships, staff apply to the institute's Scientific Promotions committee to be members of faculty. The committee includes some external members and provides feedback on applications. All fellows are seven-year appointments. The committee has recently established a Florey Fellowship as a development opportunity for outstanding SROs. This fellowship promotes them to the equivalent level of a fellow. It provides internal recognition and is tied to the recipient's existing external grant. The significance of admission to faculty is that scientists are recognised by peers as a senior researcher. Faculty meets once a quarter to provide advice to the Executive on scientific and strategic direction, and on the development of policy. While fellows need to secure their own funding through competitive NHMRC and/or ARC grants, if they do not obtain new funding the institute will fund them until this funding is secured.

Some have had a dream run in building their science research careers. One described how, by securing a fellowship, 'I am able to focus on the research and I am still excited by the science, and this has been my choice and is my preferred path' (Interview 27). Another had also fortuitously established a stellar career. The following narrative suggests several strong themes. This research fellow had good career advice, built strong international research networks and subsequently maintained international collaborations, all of which were useful in competing for funding. They also understood the importance of self-promotion:

> When I went to finish my PhD ... my supervisor said: 'why don't you go to London where I did study leave? It would be a really good place to go'. I knew it would look good on the CV but I hadn't realised when I got there how important it was. In terms of building your career and CV to have very good places where you have your postdoc or worked on your CV is incredibly important. This initial step of being in London set me up. I then met people. I went to Boston. I formed a network of people that has been very important throughout my career. Because I had worked in this lab in London a colleague back in Australia wanted me to come back and join them at a Melbourne hospital ... Making early career decisions was probably one of the most important things. Back in those days there was in the NHMRC a research fellowship stream ... that was very career orientated ... I got into that at a very early stage, came up through the levels, grew my group, maintained a lot of international collaborations. My science has always been

very international, and that has always helped. I have been on study leave overseas to good labs, built up a big network of colleagues ... You have got to learn to promote yourself. I think it is a lot tougher now for younger people starting out. I advise that unless you are doing really well you have to think if you are going to stay in research (Interview 37).

While Interviewee 37 had had a really successful career, there is acknowledgement here that circumstances are different for those who are currently moving into fellowship positions. By contrast, the narrative below of a younger research fellow has a tone of struggle and exhaustion, which is quite different to the one above where the theme of serendipity predominates:

Yes, it is my preferred career path. If you had have asked me two or three years ago I would have said a resounding 'no', because it was a pretty bleak landscape and I didn't think I could get anything up and going. Right now is good. I feel I need to promote myself more, but I don't have time ... I have so many papers that I haven't written ... I have a million other timelines I need to make. In terms of the career path, I can see that if I don't start making those applications that things will stall ... I now see that those promotions are associated with more access to other things ... I need that. I had never identified that I see that. I want that promotion for what it will do for my projects. It will be easier for me to get money. It will be easier for me to get staff and recognition (Interview 35).

Nevertheless, despite the sense of exhaustion, there was a recognition that securing promotion would open up a variety of career opportunities. Some saw science research as a tough career but were realistic about what it involved. One said: 'I can think of four times in my career where I have made an active decision to go in one direction. That's the bed, you lie in it. I don't think you can actually complain about that' (Interview 32).

Several research fellows talked about the toughness of a system where continuing in science depended on securing funding. Inability to secure funding could slow one's career and have an impact on those who depended on the funding they secured. One senior researcher explained that they had been on five-year rolling appointments their whole working life, but 'you wouldn't do that anymore, because your chances of not getting grants and fellowships have greatly increased; it is much more difficult to make a career as a research scientist' (Interview 37). A second respondent commented that 'It is not a particularly easy path. Inherently there is quite a high level of competition among scientists for resources. It is a source of frustration ... It is just a process one has to navigate'. The respondent went on to describe the arbitrariness of the funding model, discussed in chapter 1, and the impact that it had at this level:

> I ... have wondered about the arbitrary nature of some of the processes that determine who does and does not get a bundle of money to spend. Our source of income has to be won competitively. Given the rate at which those sort of applications are funded most will fail, having been involved in both sides of the process ... I haven't yet secured a substantial personal support package into the future and ... that creates some angst. Who knows, maybe the next fellowship application will get up. I make do with what's available (Interview 28).

One of the respondents quoted above talked about the impact of not securing funding on the team: 'Lack of dollars is the biggest thing in terms of keeping people on in your team. If you had a very good person but only a certain budget you might have to let them go' (Interview 37).

A fourth respondent had concerns about the transparency of the internal structure:

> Personally, I feel there is a lack of transparency and I am aware of these discrepancies. That people with strong track records are not getting as much credit for their track

records, and, vice versa, people with weak track records are getting a lot more support. There is a lack of transparency on some of these things, and it is not clear what would give you the next promotion and what would dictate your pay scale. It is a big black box (Interview 21).

A further respondent questioned the consistency of NHMRC decisions. They mentioned two applicants in the same grant round, one a male with two *Nature* papers and two NHMRC project grants (CIA (Chief Investigator A, in charge of a grant), CIB (Chief Investigator B)) who did not get a career development award (CDA) and a woman with five papers (but none higher than impact factor 5) and one grant as a CIC who received a CDA (it should be noted that the first applicant did eventually receive a CDA after being waitlisted.) This respondent believed that to be awarded a NHMRC Senior Research Fellowship 'you need to have four PhD students graduated, and more than forty papers'. But a woman who put in an application as a test run got a fellowship and a male colleague with a good track record, more papers, and more money 'was deemed as not competitive' in the next round. The frustration was that

> the goal posts keep changing. You are encouraged to publish in high-impact journals. Then the CEO of NHMRC decided to encourage translational research, and this involves medical doctors; impact factor is no longer considered, but number of publications and citation index are more important when judging track records, so they move the goal posts (Interview 33).

The McKeon Review (201, p. 32) recognised that reforms are required in the application process for NHMRC funding, asserting that 'track record definitions should be more flexible and research staff who are not chief investigators should also appear on grants to assist their career progression'.

Some of the more senior research fellows had reached a career plateau (Riordan 2011), and were no longer extremely driven, unlike younger research scientists around them. Nevertheless, the following respondent considered it was important to push their younger staff to be competitive:

> In my case I think I have found the right level for me. So I haven't sought positions such as divisional head or chased the next promotion outside the institute, just because there is more money there or it is a higher step up the tree. I think when you are younger you can be very driven. I certainly went through that phase, and that is a critical stage to go through. I see that in my young staff and I encourage them to do that, to be competitive and tick all the career point boxes (Interview 31).

Thus, having reached a plateau this senior researcher's ambitions were moderated by a sense of having found the right level and being content with their current role in the institute.

Summary

The challenges for building a career path in science research are different at each level. While most of the respondents who were doing their PhDs had not given much thought to their careers after completing a doctorate, at least one was already working with their supervisor to look at future career options. The early career researchers interviewed enjoyed their jobs and were happy with their career progression. They received support from supervisors and fellow researchers, and anticipated staying in this field. Some were already looking at the next career move, although others were not strategically planning their career paths in science research. At SRO level, respondents were on track to being independent researchers. All mentioned strong mentoring and support from supervisors—as well as hard work and long hours—as key to their success and to remaining in science research. Research fellows had mostly reached the pinnacle of their career. Nevertheless, there was the constant stress of applying every three to five years for external funding and the impact of the outcome not only on them but also on the staff in their laboratories who depended on that funding. There was evidence of generational differences here. Those research fellows now in their fifties appeared to have reached a plateau, and were content with what they had achieved, whereas younger research fellows felt the heavy weight of their responsibilities and the long hours involved in doing all that the role entailed. This generational change will be explored in detail in chapter 9. The next chapter looks at the importance of networks, mobility and mentoring in building science research careers.

CHAPTER 7

Networking, Mobility and Mentoring

Introduction
The literature indicates that networks become critical in building scientific careers, especially in gaining promotion, as discussed in chapter 2. Networking is particularly important for scientists in research institutes; for career progression they need to demonstrate community engagement and contribution to their discipline. But women scientists often find that when they have children it is too hard to write articles for publication and also do networking, so they tend to focus on publications. Others have an aversion to what they consider the political games involved in networking (Ibarra & Hunter 2007). Women in science research tend to be more embedded in formal networks, while men are better embedded in informal networks (Beaufays & Kegen 2012). Personal proximity to central men seems to be important for women's realised network position. This suggests that it is more difficult for women to access networks that include powerful male scientists and therefore women do not have equal access to science research networks. It has been demonstrated that top women scientists are aware of the capacity of networking for career development, even though still partly excluded from men's networks by power and gender stereotypes, and therefore tend to

establish and use women's networks to build their careers (Sagebiel, Hendrix & Schrettenbrunner 2011).

Mentoring is also important in building scientific careers, and can be particularly important in keeping women in science research. A meta-analysis of the efficiency of 117 mentoring programs for women's scientific careers in Germany found that they provided mentees with individual empowerment, self-confidence, and the opportunity to know role models, as well as creating interest in becoming future mentors. Moreover, the experience also raised male mentor's awareness of the situation of female scientists (Hoppel et al. 2012). The UK House of Commons Report endorsed such mentoring and asserted that higher education institutions should emphasise both male and female role models who have successfully combined a career in STEM with family life. This strategy 'could help to counter perceptions that these are women's issues rather than matters that concern all parents' (HoC 2014, p. 28). Bonetta (2010) provided examples of universities and national organisations that have created programs to formalise the process of mentoring. For example, the Association of Women in Science in the US has many chapters that bring women scientists together to network. For scientists who cannot meet in person, MentorNet (2014) is an online service that virtually connects established scientists with undergraduates and graduate students, postdocs and beginning faculty. While in Europe, the Max Planck Institute of Biophysics in Frankfurt, Germany, set up Minerva-FemmeNet (2012), a network for female scientists.

In the current study it was important to understand how important networking and mentoring were to career progression in science research and whether or not there were discernable gender differences, particularly in networking opportunities. Interviewees were asked: 'What role do different research networks play in women's and men's promotion at the institute and do women have the same opportunity as male colleagues to be introduced to these networks'? A follow-up question asked 'how important is mobility in building national and international research networks'. A second question focused on mentoring: 'What mentoring programs have been available to you at FNI? Have they been useful; if so, in what ways?'

Networking

The institute encourages networking in a variety of ways. Internal networking is facilitated through a weekly seminar series, through visiting speakers conducting seminars and lectures, through targeted seminars on career planning, and through Friday afternoon drinks at the Parkville campus. Other networking is generally facilitated by the lab head and may include introductions to their extensive networks, encouraging early career researchers to take postdocs in overseas labs and also encouraging them to present at international conferences. A particular initiative of the institute for networking and mentoring for women is the establishment of the Equality in Science committee. The committee has conducted a series of events to promote women in science (Interview 12).

All respondents in the present study considered that networking was important in building science research careers. PhD students clearly saw how networking benefitted science researchers. One commented: 'You often get PhD students coming back to the institute to progress in their career because they have built the networks during their PhD' (Interview 1). A second PhD student had observed the impact of scientists in their lab deftly networking outside the institute to get noticed, attracting postdocs to the institute and furthering their career:

> I have seen some people who come through our lab; their work is not amazing, it is good though, but they do get out there and have taken every opportunity to give talks and are always out there meeting people and they are the ones that bring in the good postdocs. It also helps to build up your CV. If there is someone coming to give the talks you will remember them. Whereas if they have been hiding away and you have never seen them before you might be less likely to give them an opportunity (Interview 2).

In addition, the student association was identified as facilitating internal networking through various social functions it convened (Interview 12).

At the next levels, for early career researchers and SROs, networking was considered 'very, very important' (Interview 15) and the role of the postdoc's immediate supervisor in providing introductions to top scientific networks was critical: 'My supervisor plays a really large role in helping me to extend my network. He actually takes me around at big conferences and introduces me to the Hollywood stars of science. I know it is not the case for everyone, and it is a shame' (Interview 5; also Interview 20).

One respondent commented that their boss was an excellent networker and that it was important for lab heads to promote staff within their networks, not compete with them:

> I have a boss who is a very good networker and who is very good to the people in his team, and tries to foster the careers of people in his team ... He promotes them within his network. There are some teams where the boss wants to take more of the glory or kudos for themselves (Interview 34).

Networks were also identified as critical to building international research collaborations, as the following respondent explained: 'You have got to be in the crowd to make your way because you need those co-authored publications in order to remain competitive and that comes from your collaborations and networks' (Interview 15).

Several senior researchers reinforced the importance of networks. One considered that the institute strongly supported its research staff to access key networks:

> It works hard to promote women and men on these networks. We work hard in grooming people to be able to be on editorial boards, various committees, put them on professional career development, in an attempt to lift their promotions so that they are more prominent and better able to be chosen for awards and future fellowships (Interview 32).

The question of any discernable gender differences in networking elicited some interesting responses. Several asserted there were no gender differences in opportunities to network (Interviews 1, 15, 18, 19, 20, 25, 27, 28, 29, 31 and 35). Another respondent had not observed any difference between women's and men's networks, although added the comment that 'it is difficult to explain the absence of women at senior level' (Interview 3).

However, others thought there were some gender differences if women had children (Interview 35). One argued that 'provided women are there for the same amount of time they get equal exposure to networks. Those who have not been there full-time will find it harder to break into these networks' (Interview 6).

A further respondent with children explained that there was no time for the social aspects of networking, confirming Ibarra and Hunter's (2007) findings:

> The problem for me is the time to actually go to these networks. My time here is very limited. I only have day care from 8 to 6. And I need to travel quite a bit to come to work. And so when I am here I need to focus. And it is hard to go and have a drink with people (Interview 16).

Some respondents did see gender differences in access to networks and networking. One argued that males tended to issue invitations to other male collaborators: 'Looking at the senior staff, the people they get from overseas to do talks, the percentage is a lot more male based. There does look like there is a bit of an old boy's network that likes to invite their friends along, particularly at senior level' (Interview 12).

Gender differences in access to networks were particularly evident at international conferences. Women did not have the same opportunities due mainly to the patterns of male socialising, an observation Sagebiel (2013) also makes:

> I think there's definitely importance in networks. It is a club, if you want to get invited to speak at an international conference. All these things then factor into scoring and assessing of your career. There's definitely a boys/girls club. If you are part of that group, by association that can help your career. If you go to conferences, they work through social interactions, going out and having a beer. That can have a gender divide … women might be a bit more apprehensive with a bunch of potential strange guys (Interview 11).

The gender differences in networking at conferences were exacerbated where the particular field was male dominated, as one younger respondent explained:

> At the international conferences we go to they are primarily men and bond with men better ... and have their annual beer drinking parties. So it is harder. That is because of the field that I am in ... When you are in the minority very much so as a female, I find it more difficult (Interview 9).

While another acknowledged that networking was harder for women, they added 'If you go to a conference run by a certain group of researchers it is hard, even as a male, to get in and talk to them' (Interview 14).

Nevertheless it was considered important for women to network—regardless of the obstacles—as an integral part of their career development, as the following respondent asserted:

> There are supporting networks [that] help you to grow your science and collaborative networks that are really the crux of good science. Men's networks are probably built around a glass of wine and a beer. It is perhaps not as easy to come in and join those social networks ... From a collaborative network point of view I think women can access the same networks but you have to put yourself out there, and look for the opportunities actively. And you must do it (Interview 23).

Thus, all respondents considered networking was important in building scientific research careers. Supervisors were highlighted as having a critical role in introducing their younger research staff to scientific networks, such as conferences, editorial boards and potential collaborations, as an essential part of career development. About a third of the respondents did not believe there were any gender differences in access to networks. Others took a different view. Some argued that it was more difficult for women with family responsibilities to find the time for the more social aspects of networking, while several thought that it was harder for women to network at conferences, especially in those research fields where men were dominant.

Mobility and Networks

How important is mobility in science research careers? In a follow-up question on networks, respondents were asked if mobility was critical in building international research networks. Zippel (2012) argues that international research and collaborations are elite activities, and that universities and funding bodies are crucial gatekeepers in this process, while González Ramos and Vergés (2012) assert that international mobility is *the* new challenge for women scientists who have to carefully plan long- and short-term decisions with regard to balancing family and work. A third of Australian SET postgrads go overseas to secure employment (Giles et al. 2009).

In this study there was a clear view that mobility was essential in building an effective career path in science research. A mid-career researcher was adamant that mobility was crucial for career development:

> Mobility is very important. You have got to go there; it has got to be face-to-face. In a big way working with other people is determined by whether or not people like you a lot. You have to be actively contributing to your field and turning up at conferences and contributing. Socially too is important, going out and talking to people. If you are collaborating with someone overseas you need to be there every six months otherwise the project is not going to go anywhere. Someone has to be pushing it. You need to be visiting and pushing the whole time for this to work (Interview 15).

The important themes of this narrative were the need to build personal and professional relationships internationally, and how this could only be achieved by being overseas frequently and working and mixing socially with collaborators and potential collaborators.

A second respondent had also benefitted from working overseas: 'I couldn't recommend it more highly. I have people in my lab and I am almost insistent that they do at least a three-month stint overseas' (Interview 21). Another argued that while mobility provided exposure to a broader world of science, it might be difficult to relocate overseas if there were children:

> It is more about getting exposure to the world's best science. A lot of people come back transformed ... It is hard to assess cause or effect. A lot of people who go overseas are more ambitious. Is it going overseas that was the key of their drive and ambition? The reality is some places overseas have a work ethic that means some people do more hours when they are overseas ... that helps them as well. I can see there are issues if you have family. A lot of people have this period overseas before they have children (Interview 27).

Most postdocs in this study had spent between two and six years outside Australia working in high profile research institutes either in the US or Europe, or in institutes in both countries. Some of those interviewed were not Australian nationals and had taken up appointments as SROs or research fellows at the institute as part of their career progression. The decision to undertake postdoctoral study in a different country to that in which they did their PhD was relatively straightforward for those respondents who were single (Interview 9) and they were usually strongly encouraged by their lab heads to do so.

Mobility was considered an important stepping stone in career progression, as discussed above. Senior research scientists could point to a range of benefits, from increased flexibility in thinking to orderly career development:

> It is just the case of being uprooted and shifting to a very different environment where there are people with perspectives that are different other than to your own. It helps to build flexibility into your thinking. On the one hand people want to see a focused effort and a thread running through career choices. They have had a sensible development to their skill set, as opposed to 'gosh they have done

a lot of different things' ... The stint overseas is given more weight potentially than being at a number of Australian labs (Interview 28).

Research scientists in the early career phase understood this requirement. One noticed that those who had done a postdoc overseas 'have had better opportunities when they return to Australia' (Interview 12), and another reported 'I have heard it said from someone quite senior that if you want to get promoted it is very good to go overseas, that you have those extra skills and having those collaborations going' (Interview 9). Interviewee 9 went further and implied there was favouritism within the institute for those who had done postdocs overseas, and this was particularly difficult for women:

> There has been some discontent of people on Parkville campus that they are recruiting people with similar skills to them into higher positions and they wonder why they haven't been promoted and that is because they have different lab experience. I think the institute look at it quite highly. I think that makes it difficult for a woman in her thirties who is thinking of having a family; you don't necessarily want to shift your life over to the [United] States for five years. If you have got a young family it is even harder. Maybe it is something we should think about in our twenties. It is easier to progress in your career as a single person in science because you can shift around (Interview 9).

Another respondent who had undertaken a postdoc overseas had received a similar message, and again acknowledged the difficulties of mobility for those with families:

> I was told before I left that it really is crucial. I have noticed coming back that the people that tend to get that next step, they have all done time overseas. That is definitely a factor. So if you had a family and moving was difficult it does hurt ... I am not sure why. Postdoc experience overseas, definitely it helps now in building networks. I was quite insulated over

there. Now I am back, it definitely makes people pay more attention to what you have done (Interview 14).

So the message—and observation—was clear. Doing a postdoc overseas was a way to get noticed and to accelerate career progression. But at the same time these respondents and Interviewee 27 above observed that it was more difficult for researchers with families to move overseas. However, the UK House of Commons Report (HoC 2014, p. 51), while acknowledging international collaboration brought benefit to science, argued that 'requiring researchers to relocate is not the only way to promote it'.

The link between mobility and funding was evident. There was a widespread belief that it was impossible to win a NHMRC fellowship without having worked in an overseas lab. Overseas research experience in the postdoc phase was essential, according to respondents, before applying for the NHMRC fellowship scheme, confirming Zippel's (2012) argument that funding bodies act as gatekeepers around mobility. But several respondents supported Acker's (2010) view that short-term placements in overseas labs of three to six months could also be beneficial. Such short-term contracts encourage mobility between institutions both nationally and internationally, foster collaboration between research groups and encourage innovation (HoC 2014). The following respondents realised that they needed to fulfil this requirement in order to progress in their career:

> One of the requirements for the fellowship above my level specifically says you have had to work overseas. It is just an essential requirement. The average postdoc needs to spend two to four years. In my career I have already moved three times to get where I am and have lived overseas. Mobility is essential (Interview 3).

> In Australia if you look at those who get CDFs [career development fellowships], the vast majority have done their postdocs overseas. The culture when I went to the US was that you earn your stripes overseas. When people are

assessing two people on who gets the fellowship, the one who went overseas is likely to get it (Interview 20).

Interestingly, the first respondent was not in a position to relocate overseas, but hoped to 'tick the box' for mobility by doing shorter stays overseas.

While the number of publications in high-impact journals was the most important factor in gaining a fellowship, reinforcing the point made by genSET (2010), career mobility was also highly regarded by reviewers and could be a decisive factor in getting a fellowship. The following story of a researcher who did not do a postdoc overseas was a cautionary tale:

> One of my close friends got married and his wife really did not want to move and he is now the same age as me and just as educated and he is constantly doing one year contracts, one year contracts, struggle, struggle, because it makes it much more difficult to get the next fellowship (Interview 14).

Mobility was difficult for the young male as well as female science researchers interviewed, which confirms other findings (HoC 2014) about younger researchers being unhappy with the system that results in a nomadic lifestyle and lack of financial stability. Several male respondents had started families while doing postdocs abroad which had an impact on their career progression. Importantly some were trying to juggle their own career progression and doing postdocs overseas with the careers of their partners. While balancing dual careers at this stage proved challenging, starting a family on top of that was even more difficult. For example, one PhD student understood the importance of doing a postdoc overseas and was already having those discussions with their partner:

> I do think mobility is important. I have had quite a few discussions with my wife and others. I think it is a huge advantage to go overseas in labs in the US or Europe. I have noticed that people who have had that opportunity

> overseas, have had better opportunities when you get back. So I am strongly considering going overseas. And thinking of the family thing, my wife has very good maternity leave arrangements, so if she wants to take time off and maybe do a little work part-time overseas. Often it is to do with the partner's interests as well. I have probably noticed more women staying here (Interview 12).

Clearly, this was a decision that the couple would make taking into account both of their careers, not just the science researcher's career, and suggested a discernible generational shift, discussed in more detail in chapter 9. It has been argued that universities need to adjust their personnel policies and working conditions to the reality of a dual career as the norm (Dubach et al. 2012), and the comments of the respondents above reinforce this point.

Ackers (2010) has asserted that the requirement for mobility is highly gendered and respondents in this study supported this view. Several female respondents had their first child when doing a PhD. While they considered that mobility was important to career progression once they became postdocs, logistically this was difficult. One had resolved the dilemma as an early career postdoc by deciding to combine clinical practice with research and not working overseas. But others had managed to move to overseas posts with their families.

Zippel (2012) argues that the biggest challenge for mobility for women scientists is juggling their partner's career with their own, rather than responsibility for children. One female respondent had taken up a two-year postdoc appointment overseas and her husband and young child accompanied her. Her husband took a lesser position in his field to allow the respondent to focus on building her career. This reflected Zippel's (2012) view that mobility can be empowering for women scientists; the notion that institutes send their best abroad assumes that they are excellent. And the positive experiences abroad can benefit women. When they return home they keep experiencing that bonus. One respondent echoed Zippel's (2012) argument, saying:

> the attitude is that anyone who comes from overseas is an expert. There is also a degree of cultural cringe in Australia.

> The effect of me doing a postdoc in the US has been more important to others than me ... It is like a box has been ticked and again I am legitimate (Interview 35).

Another had experience as a post doc in a country that had a different national context for parental leave and was empowered by knowing that career opportunities for women in science can be fostered: 'coming back I have seen how something works somewhere else, particularly on this equity and science thing' (Interview 21).

While the emphasis on mobility in building careers in science presented particular challenges for women scientists—and some male scientists—in the current study, mobility was part of a broader notion of career development. Some lab and division heads, reflecting the view that the traditional male life-course is the norm (Caprile 2012), argued that women scientists could either choose to be single and be supported in their careers or they could choose to have a family. Therefore some supervisors tended not to encourage women with children to explore the logistics of working in labs overseas.

However, a shift in attitudes to mobility was discernable. There was a view that the traditional model of two to six years working in an overseas laboratory was not the only way to tick the boxes on one's CV in relation to mobility. Ackers (2010) suggested that shorter-term appointments were increasingly more important in building a science research career. This was also considered important among respondents in this study. It was usual for postdocs to spend three months in an overseas laboratory to learn a new technique and to build research collaborations. Respondents, including laboratory heads, thought short-term international assignments were useful, and could often subsequently supplement a postdoctoral appointment overseas, but could also be used as a substitute for longer international assignments. One saw the more traditional relocation overseas for several years as outdated:

> I think the expectation that young scientists go overseas and prove themselves is a bit of an outdated way of thinking. Maybe in the past, before we had email. People are a lot more mobile, going to conferences and visiting labs overseas. I don't see that it is an absolute necessity (Interview 22).

Ackers has discussed the contributions that new forms of mobility, and particularly short-stay mobility, 'present both in terms of optimising knowledge exchange processes and "internationalisation" but also to "potentially mobile" women and men with personal and caring obligations' (Ackers 2010, p. 83). One respondent mentioned a career development seminar at the institute that had supported other models of mobility:

> We had a recent career development morning and ... that was again reiterated that overseas training is very important. It is becoming less important to do whole two and three year postdocs, but having three months training, being able to demonstrate that you can learn a new technique and bring it back is important ... it is definitely a big tick for career progression. I have not done that but I am trying to address this by organising shorter overseas stints (Interview 26).

Several senior researchers concurred, arguing that it was important to go overseas for short periods once research collaborations had been established. As one explained:

> You can build an association and put away some money to go twice a year. There are ways of doing it without pulling up stumps and going there for three years which in many ways is the best thing to do. It is not just about science it is about having a broader non-parochial view of the world ... It introduces a level of flexibility in your thinking that doesn't come from staying at home (Interview 32).

Some lab heads encouraged female staff with family responsibilities to explore alternate models of mobility, such as short-term assignments in overseas labs, and attaching a visit to a lab to an overseas conference. Thus several women with family responsibilities did manage the logistical challenge of relocating overseas, or went for a shorter period, and used this experience to leverage their career development in relation to winning competitive funding.

Mentoring

Lee et al. (2007), authors of '*Nature*'s Guide for Mentors', assert that one of the distinctive features of being a great mentor—as opposed to a great supervisor—is a special focus on helping to build your mentee's career and a long-term commitment to providing support and advice and maintaining links. They identified key personal characteristics of a good mentor as including: enthusiasm, sensitivity, the ability to appreciate mentees' individual differences, respect and unselfishness. They also highlighted qualities that mentees particularly value, such as: availability, the ability to inspire and create optimism, providing support without micro-managing, asking insightful questions while being a patient listener, being widely read and open-minded, helping to identify the right initial project, and rewarding success.

In this study, most respondents considered that the mentoring the institute provided was good, even if they had not availed themselves of mentoring programs. The culture of mentoring was inculcated early on. All postgraduate research students were offered mentors. This was organised by the postgraduate student association, as the following respondents explained: 'At the time I started at the institute they were introducing a mentoring program, but I did not have much to do with it' (Interview 6); and 'I think at the time the mentor was not available, so that was all there was to it ... the student's society at the institute is very good. They organise the mentoring system. Basically some postdocs put up their hand to mentor' (Interview 38).

The process of selection and matching was not well understood by mentees—or mentors—and appeared to operate on an informal level, but was nonetheless effective:

> ... the student's society has a mentoring program where they pair up postdocs and PhD students. And it has been really good career wise; they give you pointers on what to focus on during your PhD, how to prepare for conferences, and have been helpful in proofreading drafts. They have been really good. I have had a couple of mentors who have been really good and helpful, just in feedback about where

I am going and where I am sitting in my PhD, and what to aim for and what is achievable (Interview 1).

Others also had a positive experience and saw the role of the mentor as different from that of the supervisor; for example, 'it is a useful way of discussing problems that you have, and to plan a little for the future. That is a very good scheme. The other mentors that are available are more supervisors and they have always been very helpful' (Interview 12), and:

> The students and postdocs have an arrangement where all the students can choose to be mentored by a post doc and the senior students can mentor someone. I have a postdoc mentor and I have found this very useful to have someone outside my lab to talk to about things I am not comfortable talking about with your supervisor. It has not worked as well for some people. It depends on the mentor you have. You don't get to choose the mentor. The person is allocated to you. There is no training provided. It is a very informal thing (Interview 8).

Another respondent could also discern the benefit of having a mentor to provide advice on career development, in contrast to a supervisor whose role was different:

> When I was doing my PhD I was part of a mentoring program and that was very useful. With your supervisors you focus on your experiments and the progress of your research. It was nice to talk with someone about how to structure your career. My mentor really reinforced the importance of going overseas (Interview 14).

However, the shortcomings and missed opportunities of an informal mentoring program were clear in the account provided by the following respondent who was both mentored and offered to be a mentor:

> We do have a mentoring program. Mine didn't go very well. I got some free coffee vouchers. I think it is a great idea. I was supposed to be a mentor to somebody else. I sent them emails but they didn't get back to me. So I never met my mentee. I am not sure how we were matched. I had a mentor who was a lovely guy and very nice to talk to, but his research was in a completely different field to mine, but I could generally talk with him, but as far as our research was concerned it was like chalk and cheese. It was very mismatched. We met once ... There certainly wasn't any formal matching offered to me. It was supposed to be a very informal thing (Interview 2; see also Interview 38).

Others did not find mentoring particularly helpful. A common theme in some interviews was that the mentoring provided within the lab, often by the direct supervisor, was considered more important, even though '*Nature*'s Guide' argued that the supervisor and mentor had different roles in career development. One went as far as to suggest: 'my supervisor says he is my mentor. I don't think he would like it if I got another mentor' (Interview 18). This blurring of roles between supervisor and mentor did not necessarily benefit mentees and clearly requires some clarification within the institute. And again, two further respondents commented: 'I have a strong mentor in my supervisor. He is a very interesting character, but there is no mucking around' (Interview 19); and 'I see my lab head as a mentor. I don't see how useful mentoring programs would be, primarily because I see my lab head as such a good mentor' (Interview 29).

The strong view that the supervisor was the best—and only—mentor, presupposed that there was good communication between both parties. However, where a student or postdoc did not develop a strong relationship with their supervisor, there was the potential for their career development to be jeopardised.

Respondents considered that mentoring should be available to all research staff in the institute and that a comprehensive, formal mentoring program should be introduced, as the following respondent explained:

> There are definitely mentoring programs available ... So they are definitely available and publicised. I don't know how many people take advantage of these. It seems like something you have to make time for above your other duties. I have not been mentored but it is not through lack of opportunities. I think informally I have been both a mentor and been mentored ... I think formally there are the same opportunities offered to people, but there is in my perception a failure to take up some of these opportunities. Maybe in general it is something that could be formalised. So that if you are in a junior position you have to have a mentor (Interview 11).

This respondent was suggesting that mentoring should be mandatory for junior researchers, thereby ensuring that they were mentored in career development.

Some interviewees had not had the opportunity to have a mentor and saw this as detrimental to their career: 'Nothing, no mentoring programs' (Interview 9). Another saw significant career hurdles that resulted from not being mentored:

> I have not been introduced to any mentoring programs. This is probably why it has been such a struggle for me. I know there is one out there. I am not quite sure why I am not part of it. People here all know each other. They did their PhDs around here ... Whereas for me I didn't know what grants were there, how to apply for them. I didn't have any mentoring at all (Interview 16).

Clearly, mentoring was important for researchers who came to the institute as postdocs, especially if they came from overseas, and may consider themselves as outsiders.

This experience reinforces the benefit of the institute introducing a formal and universal mentoring program for all research staff. One respondent went as far as to describe mentoring programs at the institute as 'haphazard' and strongly argued for a more comprehensive scheme; otherwise it could not move forward:

> We need a scheme where they [the research staff] are all listed and they all get equal treatment. We now have yearly performance appraisals of our staff which means you can advise them in terms of your own experience ... What they are introducing here will involve a lot of work ... but if they keep running they will be valuable. And I am sure this is what they do in other places [institutes]. We have to do it or we will be going backwards (Interview 31).

Several senior researchers did not see opportunities for them to be mentored within the institute, but nevertheless had the resources to find the mentoring they required externally to build their careers. As one explained:

> I haven't actually had a formal mentor at the institute. But there are definitely people I have a lot of respect for. I have gone and sought people for different parts of my career. At the moment I have a PI that gives me a lot of scientific advice, and another gives me career advice (Interview 21).

A second respondent argued that the institute had only recently prioritised mentoring for more senior researchers, but they had actually been approached by a senior colleague who wanted to mentor them: 'there hasn't really been a mentoring program for postdocs and senior scientists until recently. And even then that only happened when a senior scientist approached and said he wanted to mentor me. It has always been available for students. It is very useful to get that second opinion' (Interview 15). The strong theme of these narratives of experienced researchers was the immense value of having a mentor.

The following narrative of a researcher finding a mentor is worth analysing in detail. Themes include: not being strategic about mentoring and seeking a mentor early on, engaging in informal mentoring, the importance of the institute providing an external mentor, the mentor as a sounding board and providing the big picture, and the mentor as a guide:

I didn't start seeking mentors early enough and I think they are really important. Then I did start seeking them and found it very difficult to find someone that I liked. I had informal mentoring through a couple of other senior people in the organisation who would keep an eye out for me. But there was no formal relationship. They liked me and I liked them. They would say things like 'that would be a good thing for you to do. You should think about doing that'. I have only recently been given through the institute a senior external mentor who is a senior scientist ... and it has just been wonderful. Mentors are experienced people who have been in this same career. They know where you are trying to get to. You are not in competition with them. So they can be as honest as they need to be which is just fantastic. My mentor is a fantastic sounding board for me at lots of different levels ... He has also worked in management at a high level so he can give me that big overview and help me plot my path. So I find mostly that the role of the mentor is for you to be able to say 'this is how I see things, this is where I am thinking of going, I have also got this opportunity' and they can just be someone who says 'yes', 'no', 'sounds great', 'be careful here', just give you that help and direction (Interview 23).

This respondent described what the literature now refers to as sponsorship. While mentorship has been considered important, especially for women's career development, the emphasis has now shifted to sponsorship. It has been argued that mentoring does not necessarily help women's job promotion. Rather, women require sponsorship; that is, guidance and advice, and access to key projects and assignments, in order to succeed (Morley 2013, p. 14). In other words sponsorship is a special kind of relationship and is more directed than mentoring (Ibarra et al. 2010). Sponsors are those people 'who will recommend younger colleagues, open doors for them (metaphorically), praise and encourage them, position them in terms of experiences and contacts, and help them know how to move to the next career step' (White & Bagilhole 2013, p. 184). Sponsors also introduce younger academics to scholarly networks, providing mentoring and ensuring that they make strategic career moves. Sponsorship is particularly important for women who wish to move into senior research positions. However, van den Brink (2014) cautions against the expectation that sponsorship will necessarily lead to transformational change for women in organisations; it may not challenge the status quo and may lead to those being sponsored mirroring the characteristics of their sponsors and reproducing privilege.

But some experienced researchers had simply not prioritised the task of finding a mentor or sponsor, as the following respondent explained: 'I am embarrassed to say that I still don't have a mentor. This has been on my to-do list for three years. And I mean to be a mentor as well' (Interview 10).

In summary, mentors were critical in helping early career researchers to build their careers. Most acknowledged the importance of mentoring, even if they had not availed themselves of mentoring opportunities within the institute. More senior researchers had the resources to find the mentors they needed, and often strategically looked for external mentors. Mentoring schemes within the institute tended to be *ad hoc* with no formal introduction to mentoring provided for mentors or mentees, no robust process of matching mentors and mentees, and no evaluation of the effectiveness of such schemes. Nevertheless, there was a clear view that every researcher within the institute should be provided with a mentor and that women wishing to move into senior research roles require active sponsorship. A UK House of Commons committee (2014) made a strong recommendation that all STEM employers should implement mentoring schemes for all staff, with particular attention paid to mentoring for women and other groups that are under-represented at senior levels, while an Australian report asserted that early career researchers required additional support, including mentoring (ACOLA 2012). It should be noted that since the interviews were conducted, a more comprehensive mentoring scheme has been introduced at the institute.

Gender and Mentoring

A follow-up question on mentoring in the institute asked interviewees: What strategies do you think need to be implemented at the

Florey to ensure that women undertaking science PhDs and in the early career research phase have the same mentoring and support as their male colleagues? The thinking behind this question was that women PhD students and in the early career phase may be moving into relationships and having children (Cory 2011). In these circumstances the institute could provide additional mentoring and career advice to enable them to balance work and these other responsibilities.

Some respondents thought that there was no difference in the mentoring provided for men and women (Interviews 1, 9, 10, 19 and 31). One asserted that women were more pro-active in seeking mentoring, and those that left the institute did so because of a personality clash:

> I haven't really noticed any difference between the mentoring and support that women and men get in our group. To be honest the women probably get a bit more mentoring and support because they go out looking for it, whereas some of the men do their own thing. We are a very supportive group and will help each other out, but only if you seek help. Possibly there were some women who started and left because they didn't feel supported. But it was more because they didn't ask for help and support. Sometimes it is a personality clash. If you go out and look for that support and approach it in the right way you will get support (Interview 2).

Another respondent also did not see gender as an issue in mentoring:

> I am not sure mentoring should be an issue. You should encourage both male and female scientists to find a mentor. A formal system, sure; but you have to want to do it as well. If you see it as a session every three months ... it is not going to work (Interview 10).

However, one respondent argued that women in the early career phase did require 'active' mentoring, because women tended not to return to the institute after they had children:

> Given the majority of women are going to want to have children; some active mentorship on that, what to put on their resumes, what they can do to keep the papers ticking over while they are away. There is a distinct lack of women who have had children and come back (Interview 3).

Some were interested in the mentoring program that had been instigated by the institute's Equality in Science committee: 'It would be nice to have one. I know they are setting up one; it would be nice to participate. It is the EqIS committee, it is more of a female thing' (Interview 9; also Interview 7). It is worth noting that the institute-wide mentoring program introduced in 2012 is available to both women and men.

More generally, it was considered that women with family responsibilities needed to be acknowledged and supported in the organisation, as the following respondent explained:

> I have informally been talked to by a few women and I have put my hand up for mentoring. I think it is important for women [to have] a sense that they can go and ask for advice. It doesn't even need to be a formal mentor, where there is an openness … I guess a culture where people can ask and people give, that would be good as well as a more formalised mentoring structure (Interview 23).

Thus mentoring was viewed as one important aspect of a culture of openness and transparency within the institute.

Summary

Networking, mobility and mentoring were all considered important in building science research careers. Early career researchers needed to build international networks in order to establish their reputation and get promoted. However, the 'intertwined processes of networking and gendering' (Benschop 2009, pp. 22–3) were at times evident in this case study and can 'reproduce and constitute power in action in everyday organisational life'. Thus, powerful homophilious networks were evident, particularly at international conferences and often asserted masculinity as 'care-less' authority (Grummell et al. 2009) that excluded women scientists. Mobility was an essential means of building these international networks, even though it is recognised that the ability to work overseas may be more difficult for men and particularly women with family responsibilities (HoC 2014). But short-term overseas visits or assignments were increasingly being seen as useful for career development, in particular in ticking the box in applications for competitive national funding. Mentoring both by supervisors and researchers who were not supervisors was a key to gaining the advice needed to build effective career paths in science. There was some evidence that women wishing to move into more senior roles could benefit from sponsorship, as a more directed form of mentoring. In fact, Ibarra et al. (2010) argue that without sponsorship women are not only less likely than men to be appointed to top roles but may also be more reluctant to apply for them.

While there were divergent views on whether or not the mentoring provided to women and men differed, the question about mentoring for women tended to broaden out into a wider discussion about the nature of the organisational culture and the clear barriers to women in the institute. This perception of gender differences in science research careers will be analysed in detail in the next chapter.

CHAPTER 8
Gender and Career Paths

Context
This study focuses on how women and men in a large science research institute in Australia build their careers. Most of the questions asked of the forty scientists interviewed were not specifically gender-related (see interview schedule, appendix 1). Apart from the question about whether or not women in the early career phase needed more mentoring, discussed in the previous chapter, the only other questions that had a gender dimension related to career disruption; whether or not part-time career opportunities would be useful; if in their experience at the institute gender had been a factor in career progression and, if so, could they please discuss; were there ways in which promotion criteria could be altered to ensure greater equity in the promotions process; and what initiatives did they think could be implemented to improve career progression and the position of women at the institute.

The institute has in recent years implemented various measures to address the severe under-representation of women at the level of research fellow and senior research fellow, as discussed in chapter 4. The institute's EqIS committee advises the director on strategic initiatives to ensure more effective career progression for women scientists. In addition, EqIS has recently coordinated an active mentoring program where senior people in the institute volunteer to mentor mid-career women.

All the respondents in this current study were keen to discuss gender and career paths, and many had been giving the topic a good deal of thought before they arrived at the interview. Each had their own analysis of the difficulties that women faced when combining science research and family responsibilities and suggested various strategies to enable effective work-life balance. Take for example the analysis of the following mid-career researcher about the challenges for building effective career paths for women in science research:

> I think the male career model needs to change. It is tricky because all your external funds come from people outside the institute. If they are looking at people with seven years of a track record behind them rather than five years, often they will look more favourably at people who have seven years and no breaks (Interview 12).

This narrative focuses on the male model of career progression, and the funding model to which it is tied, as weighted against women. The model assumes no career breaks and a competitive system that forces researchers to keep moving up to the next level or to exit.

However, the funding system is only one aspect of the often different treatment that women in science research may experience in building careers, and this different treatment is not necessarily related to women with family responsibilities. Caprile (2012, p. 18) concluded that 'marriage and children do not appear to have a significant influence on women's scientific productivity and academic performance' (see also McNally 2010) and that to explain gender differences in scientific careers, in particular, 'it is necessary to investigate more complex mechanisms, such as discrimination and cumulative advantage and disadvantage'. This discrimination can often be couched in terms of myths such as that males outperform females in science, and men are more prolific than women based on bibliometric indicators. Kretschmer and Kretschmer (2013, p. 34) argue that such persisting myths create 'prejudice or attitudes in relation to the assessment of women's scientific performance'.

The analysis of gender and career paths in this chapter will therefore be discussed under a number of headings that include: the absence of women; the problem is women; gender and choices; gender and organisational culture; the funding model; clinical versus research; the impact of career disruption; and strategies to improve career progression for women.

Absence of Women

The absence of women at senior levels of the institute was evident to several of the respondents. As they looked around them, senior women were in the minority. As one observed: 'I think there are a lot more political issues. Obviously a woman would be outside for a longer period. My direct evidence in my group at the same generation, there are not many women. There are still more men than women in my group' (Interview 4).

Two others observed: 'from an institute level it doesn't bode well to be seen that the entire Executive and management structure is male' (Interview 21) and 'There are large numbers of female students and post-doctoral fellows but it is not translating into senior appointments' (Interview 31). Three further respondents thought that the paucity of women at the top reflected badly on the organisation and was frankly embarrassing, and offered various explanations for women's under-representation in senior positions: 'There are a lot of things the institute could do better. It is almost embarrassing that there are so many women in PhDs and so few in top positions' (Interview 26); 'If you look at Division heads there are no women there. There seems to be a bottle neck. There are certainly plenty of fantastic women researchers out there. You hear them at talks. Maybe there is a bit of a boy's club in certain areas of research' (Interview 14).

> If you look at the higher level, group leaders, there is not one female. And I don't know if they do that actively or if women decide to spend time with their kids. I am not sure that the institute is doing that actively ('you're a woman, you can't be a department head'). Some women decide to be a smaller group leader (Interview 18).

Another argued that the absence of women was due to systemic barriers to promotion for women, across a number of countries, and considered that to succeed a woman had to make a decision not to have children or 'manage somehow':

> The main barriers to promotion are maybe the established system that is heavy to move, very men-orientated. It is a very common thing in the countries I have seen, it is mostly men. The few women that are there don't have children or they manage somehow. They are big examples for us (Interview 16).

The absence of women in senior research positions in the institute was noted by both female and male respondents and a variety of reasons presented. But ultimately they considered it reflected poorly on the organisation. The McKeon Review (2013, p. 137) linked the lack of senior women in science research to interrupted careers. But such a causal link does not explain how women without children are also excluded from senior positions in the institute, and suggests that the organisational culture of science research needs to be investigated. Recognising this under-representation of women in senior positions, the Florey has introduced the Fred P Archer Fellowship, as discussed in chapter 4. It provides salary support and $5,000 of unrestricted expenses each year for three years to a female investigator who, in the opinion of the selection panel, will most benefit from the support provided by the fellowship 'in her perceived potential to progress to senior scientific levels'.

The Problem Is Women

Morley (1994, p. 194) warns against constructing women 'as a remedial group with the emphasis on getting them into better shape in order to engage more effectively with existing structures'. Some respondents in the present study sought to rationalise the absence of women using a deficit model. They cited women's shortcomings as reasons why they did not progress in their careers. This approach focuses on women as the problem rather than focusing on 'organisational culture as the problem and take[ing] a systemic approach to re-visioning work cultures' (de Vries & Webb 2005). One identified a lack of forcefulness (Interview 7) and another saw a lack of confidence as the problem:

> Women tend not to have the confidence in science, particularly when applying for promotions. The confidence of young female researchers may be boosted by seeing women achieve at the top levels, for example Ingrid [Scheffer]. The lack of female role models in the upper echelons is having a negative effect on women in the early stages of their careers. It just appears too hard (Interview 5).

Such explanations, O'Connor (2011, p. 179) asserts: 'implicitly or explicitly define women as "the problem" and so obviate the need to look at intra-organisational culture and procedures in explaining these patterns'.

One woman researcher quoted below hinted at the negative effects of the organisational culture on women's confidence and assertiveness. This narrative is worrying. She talked initially about the importance of the current research project and how interventions are required to ensure that a generation of women is not lost to science. Then she talked about the deadening affect of the culture on her sense of achievement and self worth. It seemed that the low self esteem was a product of a culture that sees women as the problem rather than tackling the broader systemic issues that impact on women's career progression:

> I am glad this [project] is happening. I am concerned about the younger women. We need to intervene now or else we will lose this generation. We have already lost generations of these researchers. They need to know they are good and have got a voice and that their work is appreciated. I thought for a long time the work I did was shit, I was never going to get published and was not worthwhile. I no longer think that. I think some of my work is good. And I know I can get published and it is worthwhile (Interview 35).

We can see here the accumulation of even slight exclusionary practices 'that progressively disadvantage women's careers and cause a sensation of isolation, difficulty in assuming the risks inherent to the scientific career and low professional self-esteem' (Caprile 2012, p.18). Thus, the 'problem is women' attitude appears to shift the responsibility of the organisation to reflect on a culture that is not encouraging to some women to the individual, and to rationalise why women are not progressing in their careers (Ely & Meyerson 2000). It can be argued that the emphasis perhaps needs to shift instead to 'the problem' is the organisation (Fitzgerald & Wilkinson 2010, p. 136), as discussed later in this chapter.

Gender and Choices

An underlying narrative in several of the interviews in this project was that women have choices. The argument voiced was that they can decide to remain single, they can decide to remain childless or they can choose to have a partner and/or to have children. But what is missing in these accounts is any discussion of the choices that men are, or are not, required to make (Moir 2006). Thus making choices is gendered and is an integral part of being a woman in science.

As the following respondent sees it, you make choices, and there can be a sacrifice in the choice you make:

> You get women in higher positions dropping out but because they decide to have a family and have been making choices. Women have choices to be full-time and be a research scientist or they do something else. I think careers could move between part-time and full-time. But I don't see much example of that. And I think that people just choose in the end. I think it would be difficult from a personal point of view. I think it would be difficult ... I guess if people or women have been successful at juggling a career and a family, then [there needs to be] some sort of advocacy that it might be possible. From my understanding you sacrifice a lot to choose (Interview 1).

The strong theme here is that women often have to make choices and in doing so 'sacrifice a lot'.

Similarly, another respondent saw the under-representation of women as 'it just happened' and as the way science works. There was no questioning of the underlying dynamics that lead women to make these choices:

> But when you look at the heads, the overwhelming majority are male. But I don't think anybody has planned that, it just happened. Women make other choices and also it is the way the system in science works where you are judged on your publications for the last five years and your track record (Interview 2).

Another could see that the choices women made impacted on their competitiveness and that the organisation should address career interruptions, but in the end it was down to choice:

> Do we know the reasons women drop off at the top level? … There has to be concessions made for women to be able to overcome that loss of competitiveness. But it is really a matter of choice. Do I really want to work 90 hours a week running a lab? There should be initiatives like trying to deal with career disruptions. But after that it is really about individual choice (Interview 15).

A further respondent could more clearly link the choices to a masculine organisational culture —an old man's club—which required women to have a career or children, but noted that some scientists were nevertheless able to combine both:

> But it is glaringly obvious that at a more senior level that it is not the case and we discuss it regularly. There's definitely a sense of the institute as an old man's club. I have worked in three different places. Trying to take time off for children is obviously a major impact. I know there is an attitude among some of the senior people that you have to choose, you can have a career or have children but you can't have both. Some people make it work. As a younger researcher you can only hope to emulate them (Interview 3).

Some respondents argued that there must be a way in which it is possible for people's wish to have a family and combine that with a successful career could be navigated, and for that decision to be respected by colleagues within the institute. For the following respondent this was a personal concern and a key issue for career progression, which they had also discussed with their partner:

> The problem as I see it is balancing the respect for people wanting to create a family with the pressures of any institution that creates a product that has to be continually output ... Objectively you can do it, but what are the compromises that have to be made? My partner and I have spent many evenings debating this (Interview 17).

Perhaps the debate on work-life balance for women scientists and on choices needs to be revisited. As long as senior science researchers talk about choices for women as mutually exclusive, there will be little questioning of the culture that sees choices as an individual, rather than an organisational, issue. Moir (2006) asserted that science should:

> move away from a focus on personal work-life balance choices and decisions towards an understanding of how such matters become naturalised for women in science rather than men. The normative scientist as 'he' will remain in place ... unless we are able to move beyond notions of work-life balance as personal concern.

Instead, Moir (2006) argued, the discourse of scientific practice needed to 're-calibrate the work-life balance scales' by recognising how the normative male model of science is maintained.

Thus, the discussion needs to change from choices and work-life balance as a personal issue to a focus on scientific practice, definitions of scientific excellence and the assumption that a successful scientist is male (Caprile 2012). The role of the organisational culture in perpetuating this focus on work-life balance as a personal issue needs to be examined.

Gender and Organisational Culture

Science careers lack spatial and temporal boundaries (Carral et al. 2014). Hence Etzkowitz et al. (1994, p.65) argued that by accepting 'various parochial ways of conceptualising, investigating, and organising the conduct of science, significant sectors of the population have been excluded from full participation'. The concept of a successful scientist as male has ramifications for how science organisations position women as scientists, and women can often consider that they are marginalised. Careers for women can be railroaded by trying to combine family responsibilities with research, or what Barrett and Barrett (2010, p. 153) call the 'career cul-de-sac effect', born of their pragmatic early career choices, lack of access to networks (discussed in chapter 7), and both overt and covert discrimination. Caprile (2012, p. 17) talks about the 'rush hour', that is the early stage of a scientific career 'in which family and career demands most often collide, a fact that disproportionately disadvantages women'.

Moreover, a woman's career in science can also be railroaded even if she has no children. As Stark (2008, p.105) argues, the vision of women's family responsibilities as *the* possible cause of the women in science problem 'is myopic ... The singular focus on work-family balance as "the problem" is to the detriment of uncovering ... disparate treatment issues that may well induce many women to leave hostile laboratory environments regardless of motherhood status or aspirations'. Giles et al. (2009) also identified gender discrimination.

Similarly, Caprile (2012, p. 18) reported that there was no clear evidence that women without children have better career prospects 'or that they succeed in catching up with men in their careers' (see also Bagilhole & White 2013). Both Moir (2006) and Stark (2008) are clear that science must move away from personalising work-life balance as an issue only for women, to seriously examining the organisational culture and the both overt and covert discrimination they face.

It is clear that the experience of women research scientists is different from that of their male colleagues. They are often excluded from 'access to resources, influence, career opportunities and academic authority' (Morley 1999, p. 4) as well as networks that would enable them to build international research profiles. As Bell (2009, p.39) argued in the Australian context: 'due to the nature of scientific research as a male-dominated sphere, the work environment, the lack of role models and mentors, and the gendered notions of merit and promotion all arguably have a detrimental effect on the advancement of women in science'.

It was therefore not surprising that Asmar (1999, p. 269) found that women in the sciences in her Australian study were 'more isolated, less secure in employment, and less involved in and/or committed to their research'.

How then did respondents in the current study view gender and organisational culture in the institute? Many were keen to discuss the topic. As one explained, 'gender differences are probably related to the culture. The aggression (competition) is probably a male thing' (Interview 4). Others were clear that men's homophilous networks (Grummell et al. 2009) were an obstacle in career progression; for example, one said: 'People talk about cultural fit. So that grey area of cultural fit could be about [how] they fit that gender. It definitely could be a factor' (Interview 11). But they also described overt discrimination in the form of sexism that Stark (2008) identified as intentional sex discrimination in science laboratories:

> I have been here long enough to see women and some men fall through the leaky pipeline. I think our male leaders favour people who look like them. Hence, young men are promoted ... My lab head does not discriminate. However, there are some labs heads who are openly sexist and make comments about clothing and joke about spiking coffee with contraceptives (Interview 5).

Curiously, these male lab heads—jokes aside—were perpetuating a culture that asserts women researchers have to remain single in order to stay in science research. Such homophilious networks were also identified by another respondent, who saw this as a generational issue with older men in positions of power that they did not want to relinquish:

> Understanding what's causing underrepresentation of senior women is the first point. I guess there is an historical remnant ... Specifically for women ... what do you do? I don't know. I am in a pretty male dominated little group ... You almost have to wait for people to leave or retire, because you can have your own little empire and until that person retires they can largely control their empire (Interview 11).

This boy's club appeared to be entrenched and controlling; hence the reference to 'control their empire'. It was also described by several respondents as competitive, aggressive and a difficult climate in which to operate. It was compounded by the job insecurity with even senior researchers surviving on short-term competitive funding. The following respondent painted a rather unpleasant picture of how this culture operated, reinforcing Stark's (2008) findings, and went as far as to describe it as at times 'nasty':

> Science is a boy's club, always been a boy's club; hopefully it won't be, but it is at the moment. The majority of people in senior positions and even moderately senior positions are men. They have a nature of relating to their students and their juniors which in the end has more to do with a football club rather than most well organised workplaces. That is contributed to by the fact that there is a competitive element to scientific functioning both in terms of the actual output ... but also in the basic interactions, the meetings, the talks, etc. In the end they have an element of challenge and conflict to them. That gets occasionally nasty, aggressive and unpleasant. And people in that situation use all of the tools available to them, including the more unpleasant ones. It can be a very hostile environment, and compounded with that you are surrounded by people who are incredibly insecure in terms of their employment, in terms of their life. They really are competing for the basics of their employment. In that kind of situation people will use whatever tools that are available (Interview 25).

This account suggested that in such a competitive environment any sense of cooperation and mentoring was subsumed by a race to progress, or at least survive, in one's career.

The effects of an organisational culture that talks about choice for women as career or family, and where processes around promotion are often perceived as not transparent, is to produce a sense of *ennui* for women in the institute who are trying to forge ahead with their careers. The following narrative recognises both the formal and informal barriers that women scientists still experience, especially when they reach mid-career:

> There's definitely more increased awareness particularly at the NHMRC level about career interruptions and that is definitely meant to be taken into account when people are making reviews. But I am also very conscious of different attitudes towards people in science and some say 'yes, when I pick up an application I *very* much take on board when an individual says I have had a leave of absence for whatever reason and take that into account' when they're reviewing, but there are many others that just do the maths and say 'you have been in the game for ten years and you have published twenty papers. Yep that's okay'. Or 'you have been in the job for ten years and have published ten papers; that's not good but oh, I did not give you any credits for having a leave of absence'. So I am very conscious of that, while we're all meant to be recognising career interruptions, how much it is really going on. And there's the old school of thought and the new school of thought. I have heard people say in *this* building from our senior levels of staff that a woman has a choice, she can have a career or she can have a family. That can ring pretty heavy in your ears when you are trying to set up an independent lab. And I am trying to be very strategic (Interview 21).

The discussion in this narrative about the randomness of how funding bodies assess output relative to opportunity and the belief of some senior scientists about women having choices to be a successful scientist or a parent, made it difficult for women to push forward in their careers and be promoted to senior positions.

The highly competitive environment meant there was little consideration of women who wanted to have children or men who wanted to work less than full-time in order to help care for their children. One respondent talked about how this culture needed to change and mentioned the even more rigid attitude of a laboratory overseas towards women having children:

> If there is any prejudice it would be around the fact of having kids ... In the lab I was in overseas, it was terrible. There was a girl who had a baby, and the professor there basically told her not to come back. I thought that was pretty terrible (Interview 14).

In these circumstances it was difficult for men and women to negotiate part-time work because science careers are generally predicated on working full-time and long working weeks: 'but being half time you are half as productive. So if you are to be part-time and you stress out about being not as productive then it doesn't work', and, being part-time, 'you are against a system that is only fulltime' (Interview 16).

The organisational culture described here is tough, extremely competitive and even nasty, and overtly gendered. It is not for the faint-hearted. It requires a strong commitment to one's career to the exclusion of any other responsibilities and social relationships (Carral et al. 2014). It is clear that it can be difficult for researchers to negotiate any flexibility in working conditions even though the national Fair Work legislation makes provision for flexible work arrangements for parents, and parents may request a change in their working arrangements (see White 2011).

Funding Model

As discussed in chapter 1, the funding model in Australian science research shapes the nature of science, the concept of excellence in science and the notion of a successful scientist. Consequently it is difficult for women with family responsibilities to compete with males. While the NHMRC has recently introduced assessment of funding applications to take into account output relative to opportunity, few respondents believed that this was seriously assessed. Moreover, as the following respondent who has been an assessor of NHMRC applications explained, there are no effective guidelines or training for assessors on how to assess output relative to opportunity:

> Panel members are not trained at all. I have been on panels. So I can tell you there is no training around what you call a career interruption. I have not heard any derogatory comments about applications that have had career interruptions, but I have spoken to a person who has been extremely scathing about some things that people put into their career disruption segments. On the other hand I know there are women here who this year had to put in the birth certificates of their children and provide detailed information about how long they had off ... really intrusive documentation which I think is ridiculous ... So I don't think it is taken very seriously (Interview 23).

The funding model therefore continues to see the traditional monastic male scientist as the norm and according to some respondents does not consider women with career interruptions as serious competitors for the limited funding available for science research. The reality is that men with better track records and usually an uninterrupted career will secure the funding, as one respondent explains:

> The moment you take time out of science, even with the NHMRC saying productivity relative to opportunity, but since it is not written anywhere, [it is] how to work that out. In the end if you are faced with person X with 20 papers and person Y with five papers, on the whole in the competitive funding environment you are going to go with person X and say that they have done better. It is perfectly possible that a career break means you are never going to catch up (Interview 40).

Part of the problem, as another respondent highlighted, was the small funding pot, and the other part was what new models might be required if the NHMRC moves away from its usual practice of funding productive male scientists:

> The NHMRC are not a long way behind in having heard this a million times ... The problem they have, this is my belief, is that with a defined pot from which to fund research, they find it very difficult to tell the monastic male that has ticked all the boxes, that his superb research is not worthy of funding because they have to divert money to ensure there is a labour force participation ... that is going to reap productivity for the nation. The real answer is that Australia does not invest as a percentage of its GDP, an appropriate amount of money so that the monastic male is not punished (Interview 36).

Scarce financial resources therefore lead funding bodies to pick winners based on strong research output in high impact journals. However, it should be acknowledged that the current funding system may not necessarily fund the brightest and best scientists, or the best science, given that the review process has serious shortcomings, as identified by Graves et al. (2011). The model also puts enormous pressure on applicants. Herbert et al. (2014) found that the huge amount of time involved in preparing grant applications was stressful and conflicted with family responsibilities; it restricted family holidays and led to work being done on applications at home which impacted on partners, family and friends.

The UK House of Commons committee has questioned this type of funding model and has recommended that fellowships be advertised with the option of working part-time. It also identified a lack of coordination and communication between funding bodies and higher education institutions which, 'exacerbated by the use of short term contracts, results in women falling into cracks in the funding system when maternity support is required' (HoC 2014, p. 39). In addition, it recommended that funding bodies make their maternity leave provisions clearer to both researchers and employers.

Clinical versus Research

The following discussion of clinical versus research does not identify individual interviewees by a number in order to ensure anonymity. One of the interesting observations of this project, in looking at gender and career paths, is that there appeared to be more sense of collaboration, rather than competition, for the institute's research scientists who are based at the Austin Hospital site and combine research with clinical practice. Moreover, women found this model more inclusive and more attractive in building science careers. Some had moved to careers in clinical practice and research after they experienced lack of mentoring and career support in their early postdoc phase. Several respondents noted there were more senior women at the Austin campus compared to the Parkville campus, and consequently there were some good female role models at the Austin. Others argued there was more job security because those combining clinical and research earn considerably more from their clinical work than their research. A further attraction of combining clinical practice and research was that research tended to be translational, which was now a focus of some external funding bodies. Certainly the recent McKeon Review of the NHMRC (2013) supported translational research. However, several respondents talked about the challenge of balancing clinical practice and research.

In summary, women were often attracted to combining research and clinical practice because there was more financial stability, but also because the number of women in senior roles at the Austin produced a collaborative leadership model. This was more appealing for some respondents than the perception of aggression or even 'nastiness' observed in the laboratories at the Parkville site and discussed above, under the heading 'gender and organisational culture'.

Impact of Career Disruption

So what is the cumulative effect of a competitive, even aggressive, research culture on women who have had career disruptions, and secondly, what is the impact of career disruption on the careers of women scientists? The degree of support provided to a woman who steps out of the lab for a period of time to have a child can be critical in determining the impact. The following narrative from a lab head explains the effect of career disruption and returning to work part-time on a woman postdoc they supervised:

> My most direct experience is with my post doctoral fellow … The birth of the child took her out of the scene for a while and she is now working part-time … Particularly at the time she left, a number of opportunities fell by the wayside. It had been a particularly fertile time in terms of data and analysis. There were a few things earmarked for her with the idea of cementing her profile but then she was gone and the responsibility for doing those things had to be moved to other people. Her position in the list of authors changed as a consequence. So that was an unavoidable negative impact. If she had given birth a few months later it might have been a different story. Now her productivity reflects the amount of time she can feasibly be at work. She can absolutely tick the box for career interruptions due to child care responsibilities. But it is not just that. It means that there is a bunch of things that she might have contemplated, like moving to a lab overseas, which we were trying to arrange before she fell pregnant that may not happen now. A few years elapse and you are not in the right kind of selection criteria for some of those sorts of trips. I can see how it would have a very substantial impact and that is a story that is presumably replicated a number of times for women (Interview 28).

One questions why this high performing woman scientist was not kept linked into the projects on which she had worked during her maternity leave, why an RA was not employed to keep the project ticking over, and why the supervisor had not planned a strategy for maternity leave and transitioning back to work after the birth of her child. These issues will be considered in chapter 10. There is a sense of inevitability in this narrative about choices the woman made—to have a child—and consequences of those choices.

When the PI sent the above paragraph to the interviewee for permission to use the quote from their interview in context, they responded with the following comments which focus on the tough, competitive environment in which postdocs work. In these circumstances, even though the supervisor provided mentoring and support, peer review by grant and fellowship panels does not necessarily look favourably on 'gaps in track record':

> The project was part of NHMRC funded work of which the postdoc's contribution was part of a team effort. As a chief investigator on the grant it was my responsibility to deliver the project. The key performance indicator of grants is publication. The postdoc was an author on the publications that ensued and which she sighted while on maternity leave lasting 12 months. The point I made in the quote was that the postdoc wasn't the first author of the papers: 'Her position in the list of authors changed as a consequence'. First authorship is the goal of postdocs interested in career advancement. However, you can't be a first author if you don't write the manuscript and do all the intellectual heavy lifting that first authorship implies.
>
> I take an active role in advising, assisting and promoting all my students' and fellow scientists' careers. Unfortunately, this advice and assistance extends only so far. Critical aspects of career development are usually determined by peer review (grant and fellowship awarding, publication). Review is at arms distance and so the people who know the applicants best of all are usually excluded from the decision-making process lest conflict of interest occurs. Grant and fellowship panels try to pick a precious few winners from a large, high quality field. Any gaps in track record are used to whittle down the field. This is the environment in which a path interrupted by childbirth is appraised. The very best co-managed plan between post doc and mentor will not necessarily succeed in this environment.

Another supervisor told a story of his experience of working with female colleagues who took maternity leave. There was a sense that going off and having a child begged the question of whether or not 'they were in the right area in the first place'. The corollary of this thinking would be that if a woman scientist wishes to have a child she is not in the right career. However, the supervisor could at another level see the inequity of this choice and therefore supported the research project:

> I have had enough female students and staff within my team to know that people can be excellent students, they can undertake postdoctoral training and then they can adopt a family life and not take up the challenge of re-entering the scientific workforce. There have been several who have given it away entirely or gone into industry. However, in some cases it may have been that they were not in the right area in the first place ... But those who have the drive and the intellect and can manage thinking for themselves, and can write grants, etc., then obviously there is going to be a frustration point when they are dropping behind (their peers) depending on the number of children they have or the amount of maternity leave they take. I am well aware of the impact this can have and that is why I am supporting this sort of initiative (Interview 31).

Thus while supervisors we aware of the adverse impact that career interruptions for women and men could have on career progression, they were working within a highly competitive funding system where scientists are rewarded on output in high-quality journals. Even though they were supporting female postdocs who took maternity leave and then returned to the lab, the lower productivity of these women compared to colleagues who did not have career disruption would inevitably slow their career progression.

Strategies
Respondents were keen to discuss strategies that would keep women in science and, in particular, strategies that would enable women with family responsibilities to effectively build their careers. Some talked about affirmative action initiatives as one way of addressing the paucity of women in senior research positions. There was acknowledgement that the institute (Interview 37) and also the major funding body, the NHMRC, had already implemented some

strategies. The institute's EqIS committee was singled out for mention as a strategy that was changing the organisational culture: 'I think this is where EqIS has been important, it creates a culture in an organisation. My colleagues here go to meetings of that group and will talk about it over lunch. So it becomes part of a culture and is useful in that respect' (Interview 17; also Interview 27).

But there was a sense that a good deal more needed to be implemented. In the UK, for example, it has been recommended that diversity and equality training, including unconscious bias training, should be provided to all postgraduate students. In addition, such training should be mandatory for all members of recruitment and promotion panels for STEM jobs in universities, and line managers and supervisors of staff (HoC 2014). The following narrative acknowledges achievements to date, but sees the next challenge as increasing the representation of women as senior research fellows. At the same time, the intense competition was also acknowledged as well as the impact of homophilious networks:

> The strategies in place at the moment are reasonable. Even the NHMRC has taken some steps. What Florey needs is some strong senior females who run labs, because you always need people to look up to. And they don't have them now. That is the biggest driver. Previously there was only one senior female. That one person left, so bringing someone in maybe is very good. The people [women] they have coming through are very strong. A lot has been done. There are opportunities, fellowships and the like for women. Getting over the hurdle to senior research fellow is a big hurdle. Support from above, outside the institute, needs to be bigger. That is word of mouth. The problem is that it is very competitive and senior people at the institute may have their own people they want to promote (Interview 19).

Another respondent thought aloud about what the problems were and the impact this had on women's careers in science:

> There has to be a cultural understanding and I perceive at my level that will be achieved. If a female scientist chooses

to have children it is going to impact their productivity. It is going to be tough. The timing is crucial. Maybe you need separate schemes. Maybe you need affirmative action. I basically put myself in my wife's shoes. If I had children it would have taken me off the map in science. It is right at that point [when] you are trying to become independent. The way to get around it is productivity relative to opportunity. It would be better to say that it took you out for two years. Three months, no way. Do you have affirmative action? There are 65 grants to hand out; maybe half of these are for females (Interview 20).

The perception here was that women often had children at a critical point in their career trajectory where taking time out could take them 'off the map in science'. Therefore, affirmative action initiatives might be required, such as earmarking a number of grants for female applicants with career interruptions.

This point was further discussed by several respondents who suggested strategies such as the NHMRC and the institute developing new funding schemes that were specifically for women with career interruptions. One thought such a funding scheme should be available to anyone who is unable to work full-time:

I think at the Career Development Award they need to develop a scheme. That is the time at which women drop out of the system. It needs to be strategically a separate stream. Then you are competing against other people that have career interruptions. In an ideal scenario it needs to come at a government level. The NHMRC needs to be introducing a scheme for people who are unable for whatever reason to work fulltime (Interview 21).

Another respondent focused on the institute providing more 'female specific information sessions and the like' (Interview 26). Reflecting the views of the above respondents, the UK House of Commons report (2014) has called for a change in workplace attitudes towards maternity leave and for academia to address the real and perceived career damage that can be caused by taking parental leave.

Others thought that the institute needed to take more initiative, and saw the desire to work part-time as a couple, rather than a specifically female, issue: 'Availability of child-care is a key issue. It is about choice. The reality is that some women and men are going to want to go part-time for some time. These are some of the issues that need to be addressed' (Interview 27).

One specific initiative suggested was for the institute to support women who were rising stars but whom the funding system had failed. The respondent talked about the unfairness of the current system for women scientists:

> One big one would be for the institute to acknowledge females who have been let down by the NHMRC, ARC grant fellowship system. There are women who are brilliant scientists and have had career disruption and this is not fully taken into account when they go for these applications. The criteria for these schemes are dependent on how many years post-PhD you are. You can have a male who has had a productive year every year, as against women who have had career disruptions. I don't really think that is fair. So without trying to change the NHMRC system the Florey would offer fellowships for these women who have been left off. Practically, it is about setting funding aside—whether a percentage of grant money—so that these particular women don't fall through the cracks (Interview 26; also Interview 28).

The unstated assumption here was that science is currently losing really talented women researchers and that the institute, and the nation, cannot afford the continued leakage, a sentiment echoed by the UK House of Commons (HoC 2014).

Other suggested strategies by respondents were much more targeted. One was for the institute to offer women career planning when they were thinking about having children or were first pregnant, rather than targeting women at a particular point in their career trajectory. Another strategy was for women to be more proactive about the career choices they have made. Several discussed the barrier of working part-time, because men were generally not required to make this choice and because part-time was often considered to signal less commitment to one's career (Ceci & Williams 2009), rather than a sensible way of managing and balancing work and family. The following respondent had decided to tackle this discrimination, including that from other women, head on to overcome part-time researchers being regarded as 'invisible' in the organisation. Her experience is reflected in Dubach et al.'s (2012) research that found academics who were mothers were four times more likely than fathers to feel they were no longer taken as seriously or supported as well in the workplace since having children:

> There is a very significant problem with people who have career interruptions and people who work part-time. Part-time people are invisible. It is just so peculiar; you might be doing the same quality of research and publishing in the same journal but not be regarded as serious about what you do if you don't work five days a week. It doesn't matter what you do.
>
> It is not just the perception of male colleagues [about part-time equals less commitment]. When I was training, a colleague had a child and she came back to work. There were quite a few nights her child was sick and she would ask me to cover for her. I would remember thinking 'bloody X'. We judge other women terribly. Junior colleagues look at you when you say you work part-time and there is a sort of a shudder.
>
> I am making it a political gesture ... I make it clear when people ask me, that I work part-time. There is mental balance and you don't have to have a justification. It is not a sin (Interview 35).

This narrative also suggests that other women, as well as men, can be negative about part-time work and about women needing on occasion to prioritise children over work. One lab head argued strongly that their colleagues should focus more on keeping women's careers on track when they are working part-time: 'I don't think a lot of lab heads are sensitive to it or really appreciate the fact that if you have someone who is .5 and you keep them on a project as an equal chief investigator and it keeps ticking over, then the publications keep coming out' (Interview 37).

What is different in this account is how the lab head sensibly looked for strategies to keep the research progressing, rather than considering that a woman's choice led to inevitable under-performance in the highly competitive funding environment and possible exit from a science research career.

Finally, the current trend for more women than men to enrol in PhDs at the institute, discussed in chapter 4, provided another reason to develop strategies to retain them in science and prevent the leaky pipeline occurring on the Parkville campus, where there were few women as research fellows. One respondent argued strongly that the institute needed to be creative in developing strategies to retain women research scientists because they were more likely than men to remain in science (Interview 37), reflecting a reality on which governments in the EU are now focused (genSET 2010). Another said that the institute 'can't afford to just let them go', and 'if women are lost, it is a loss to the institute and to research on the whole. We have got even more than 50 per cent of women doing biological sciences but we lose them. It is society as a whole, how do we get work-life balance?' (Interview 34). So the attrition of women from science research was considered a much bigger problem than the impact on the immediate work environment. It was a loss not just to science but to society 'as a whole'.

This respondent reflected on the existing organisational culture and the double standards applied to men and women in science. Men are not supposed to be balancing work and family (as Moir (2006) emphasises), and when women do so they are 'seen as lesser':

> A man that is trying to balance that as well is looked down on, which is really interesting. Yes I think that organisationally that is looked down on, they're not as driven maybe. I think that that is really important too. That is the other side of this coin ... A woman is seen as lesser because she is not going to be able to put that intensity in. But he might not want to put that intensity in, if he wants to actually have, as society is realising it is important for the man

to have, a nurturing role too and be involved with the family, and that the 1950s model was not that mentally healthy (Interview 34).

This narrative argued that the current stereotypes of men and women in science were outdated and that the construct of masculinity on which science excellence is based is outmoded, consistent with Etzkowitz et al.'s (1994) assertion. It also argued that new models needed to emerge and were in fact emerging.

Summary

This chapter has explored how gender can be a factor in building career paths in science research. It examined a number of themes. First, the underrepresentation of women in senior ranks at the institute was noted by both women and men who considered it reflected poorly on the organisation. Second, discussion of women as 'the problem' shifted the responsibility of the organisation to reflect on a culture that is not encouraging to some women back to the individual. Third, the view that women can choose to remain single, to remain childless to have a partner and/or to have children and that these choices will impact on their career progression underlines the role of the organisational culture in perpetuating the focus on work-life balance as a personal rather than an institutional issue. Fourth, the current funding model makes it difficult for women with family responsibilities to compete with males, and for supervisors to support women with career disruptions because they were working within a highly competitive funding system. Fifth, women were often attracted to combining research and clinical practice because there was more financial stability, but also because senior women in these roles produced a collaborative leadership model. Finally, respondents outlined a range of strategies that would enable those with family responsibilities to effectively build their careers. These strategies indicated that significant change was needed in order to keep women in science.

If the institute is any indication, new models are emerging in science research and these will challenge how research institutes in Australia and internationally do science and challenge, in particular, an outdated funding model. The next chapter will explore the significant generational change that is already occurring in science research and its implications for gender and career paths and, just as importantly, for the operation of science research institutes.

CHAPTER 9
Generational Change in Science Research

Generations in the Workplace

Many of the science leaders in scientific research centres are now in their fifties or early sixties. They provide leadership at an executive level but also in the research laboratories. But a new generation of scientists is emerging that has experienced different influences in building their career paths. Erwin Neher, the Nobel Prize winner for Physiology/Medicine, explained that the problem for Gen Y as young investigators in Europe has been that they could not follow up their own ideas at an age when they are most productive (*Euronews* 2012). This indicates that the structure of science research careers produces frustrations for young scientists, and implies a tension between Gen Y and the older generation of scientists. There are currently four if not five generations in the workplace. In the US the median age of the working population is around 36.7 years, but in emerging markets it is closer to twenty-six. The younger generations, according to Shah (2011), have very different workforce behaviour and expect their work environment 'to allow or even encourage them to use' collaborative tools. Shah also asserts that given the overlap of generations, it is not simply younger workers being mentored by older colleagues, 'but them learning from other peers in other areas and younger folks how things may work differently in different environments'. In other words this collaboration needs to be spread across the organisation.

The current study did not initially focus on generational change in science research. However, early on in conducting the interviews the PI became aware of a significant shift in attitudes about how one does science and the structure of science careers. Therefore the PI included an additional question to those listed in the interview schedule asking interviewees if they considered there was generational change occurring at the institute. The findings in this study resonate with a recent survey conducted by the Association for Women in Science (AWIS 2012). The survey of 4,225 publishing scientists and researchers worldwide found that that lack of flexibility in the workplace, dissatisfaction with career development opportunities and low salaries were driving both men and women to re-consider their profession and this dissatisfaction was more pronounced among scientists under forty.

In this study there was a strong response to the question about generational change. The issue, as one respondent identified it, was that: 'there is a significant difference in the generational gaps of the thirties and fifties' (Interview 32) and another asserted: 'Gen Ys are the ones who will want to do it differently. I have got Gen Xs who are just as hardened as the Baby Boomers. These Gen Y guys we can change' (Interview 35), while a third said: 'The generation now, probably have very different views to the generation that went before. I would like to think that my friends and I would see ourselves as equals; whether or not we are men or women' (Interview 8).

The shift was most often articulated by younger male research scientists. One respondent, in particular, encapsulated the shift in thinking between the two generations:

> I am very pleased that this [research project] is being brought up. This is not just a woman's issue, it is a family issue. And that impacts a male as much as a female. I hope it has a hugely positively impact for women thinking about these things. But I also hope it helps males as well. Looking at other males, we like to spend time with our families (Interview 12).

These comments suggest some important themes. One is that younger male scientists consider that family issues for both women and men are not being addressed in the current science career model. Another is that men in science wish to achieve more work-life balance by spending more time with their families. This thinking represents a seismic shift from an earlier generation where men were assumed to be single-mindedly devoted to science or prioritised their careers at the expense of their female partners. This research confirms the findings of the AWIS (2012, p. 6) survey that showed a significant difference in attitudes to work-life balance or integration between younger women and men and older scientists. For example, in the AWIS survey 58 per cent of those aged thirty-six to fifty-five said that work demands conflicted with their personal lives at least two to three times per week, whereas 43 per cent of those aged fifty-six and over experienced such conflict. The generational difference is also confirmed by the UK House of Commons report where the Russell Group Equality Forum asserted that 'caring for family members is increasingly becoming an issue for men as well as women'. It argued that this change should be recognised and that men should be facilitated to play active roles as carers for both children and elders (HoC 2014, p. 37).

This chapter will examine how the current generation of science leaders have determined the way in which scientists need to do science. It will then analyse how the younger generation are challenging the old model of doing science, and how they are thinking about more flexible work models and are focussing on achieving work-life balance.

Impact of Baby Boomers on Doing Science

The Baby Boomers are those Australians born between 1946 and 1965, a period characterised by high fertility in the Australian population, after which the fertility rates declined (*ABS* 2013). Gen Xs are defined as those Australians born between the early 1960s and 1982; whereas Gen Ys are generally defined as those born between the late 1970s or early 1980s and the early 2000s. Typically then, Gen Xs would be in their early forties, and Gen Ys in science research would now be between their mid-twenties and mid-thirties.

It was clear in the interviews conducted with forty scientists at the institute that there were marked differences in attitudes of the Baby Boomers on the one hand and Gen Xs and Gen Ys on the other. Take, for example, the point of view of one Baby Boomer in relation to the next generation:

> This is in part again I think a generational thing ... I have the impression that the current crop of postdocs are far less able to reference themselves in terms of understanding where they sit in the world and reference much more other points ... If you have a lab that has got more women in it, that sometimes gets accentuated (Interview 32).

Another complained that new technology was distracting younger researchers from doing science: 'The sheer inundation of data via the internet; people are not doing enough science reading. Time to do quiet reading and research is all gobbled up these days. And furthermore, the bar keeps being raised, so it is harder and harder to be competitive' (Interview 31).

But most of the comments about the difference between the generations came from Gen Xs and Gen Ys. And the common theme was that it was time for the Baby Boomers to let go of their hold on power and allow the next generation of science researchers to shape the way they do science. The following comment neatly states both the problem and the solution:

> I was answering these questions [in the interview schedule] to myself last night. I wrote 'make the boomers retire at 65'. In reality what needs to happen if an institute wants to have women as a presence, and they should because our game here is we want a number of different perspectives to solve issues. So you want to recruit from a variety of different locations around the world or experiences, and you want to recruit a variety of genders. So then you can get the best triangulation and ideas to solve problems. So it is in an institute's interest to have diversity (Interview 20).

This respondent was arguing that until the current crop of leaders retired, the institute would not achieve more diversity, including more women in senior positions. Others had a similar view: 'people are going to have to drop off their perch before the people below are going to take over' (Interview 25). Another respondent asserted that: 'One of the big issues in representation [of women] is that the people sitting at the top are at the top and are not going anywhere until they retire. So I think that is sitting on top of the data for all young people' (Interview 11), while a third respondent said: 'I would not say the Florey has been pro-active in getting people as high as possible, and that is because people above are trying to stay there; the NHMRC is a competitive system at every level' (Interview 19). The resounding message then was that the Baby Boomers needed to let go and to retire.

The consequence of the Baby Boomer generation remaining in power was a narrow model of how careers are organised, and of the position of women in science. But, according to one respondent, this was changing:

> A man in their sixties has ingrained in their bones their superiority. They have been a primary 1950s husband for the majority of their life. That is how they work. That is how they act. That is not true of someone of my age ... [in medicine and science]. The overall weight of women being in a position of seniority is rapidly increasing (Interview 25).

Another saw Baby Boomers as having simply drifted into science research and not being strategic in career planning:

> Did lab heads a generation ago also work long hours? I don't know. Essentially, there is a large gluttony of people who go into science as a path of least resistance, because they are not really trained for anything else after completing a PhD, so they go into science research (Interview 26).

A further respondent saw the 'old model' of career and family as a thing of the past and that Baby Boomers did not understand the demands of women and men as Gen Xs and Gen Ys trying to combine career and family:

> It is possibly a generational thing. It has to be a gender thing too … A lot of guys as lab heads probably are lovely guys … but they grew up in the generation where they worked all the time and their wives stayed at home with the kids and don't realise the demands that can be on younger people looking after families (Interview 10).

The consequence of this older way of thinking on Gen Xs and Gen Ys was evident in the interviews. The following narrative continually weaves between describing the behaviour and demands of the older generation, and the demands and preferences of the younger generations:

> My lab head believes it is essential to go overseas for a period of three years or more in order to further my career in science. At the moment I am dealing with extreme anxiety about this, as my partner and I do not have the flexibility to leave Melbourne for the next couple of years. This linear and restrictive career trajectory is endorsed throughout the Florey. A lot of men of my generation want to take a more care-giving role, and have partners who are professionals and cannot move overseas with them. In our lab we are seeing this change; my male colleagues are bringing in their children and expressing their angst over their partner's careers not being compatible with overseas travel. There is a generational shift and people at the higher levels need to see this. I have heard that on promotions committee they ask what your husband does … There is evidence of outdated thinking at the top. We really need to see this change, to make this career a little more sensible for the reasonable person, with family or a two-body problem (Interview 5).

While the old thinking caused anxiety for this younger scientist, at the same time there was evidence that the younger generation were redefining the workplace (younger men bringing their children to work) and in doing so redefining the traditional male model of an excellent scientist and what sort of scientific knowledge is valued.

Again another respondent moved between describing the attitudes of the older generation and those under 50 who understand that women now have careers:

> It tends to be the older generation of men that are far less receptive to the importance of recognising these career interruptions. People under fifty are far more receptive to these career interruptions and balance at home, and that women have careers now. It depends on your lab heads; there are young guys that are aware of that [evidence of a generational shift]. It is great to know that we are changing (Interview 21).

The Gen Xs and Gen Ys were optimistic about change in their workplace. And part of that optimism included a belief that more women would move into leadership positions. The following respondent also emphasised the greater discrepancy in gender representation in leadership positions between the institute's Austin campus, which combined clinical practice with research, and the Parkville campus, which engaged in basic research:

> I think it is noticeable that at Parkville all the lab heads are male. Maybe that is a generational thing. Those attitudes may change in fifteen years' time when the current crop of women postdocs gets to those positions. We have more equal representation here at the Austin. This is more clinically based and more of the senior women here have a clinical background. At Parkville there are women postdocs who are really good. But the lab heads are men (Interview 10).

Other Gen Xs and Gen Ys were calling for more widespread change in how Baby Boomers do science. Take, for example, this respondent's call for young scientists to start agitating for new ways of communicating science using new technology and for a less hierarchical structure:

> To a certain extent the young scientists should be agitating to get to the top. I need a virtual presence. I need my information on the net. The older guys don't get that ... I don't know if the 'powers that be' appreciate the internet as a virtual space. In some ways the Florey is a bit more militant in its structure than other institutes. Division heads, lab heads. There is this extra layer of management (Interview 20).

This fundamental shift in the way young researchers communicate science has been noted by van Noorden (2014) in a *Nature* survey of 3,500 scientists and engineers in ninety-five countries. The survey found that social networks such as Google Scholar and ResearchGate were widely used and provided a place to create profile pages, share papers, track views and downloads, and discuss research.

Younger researchers were also calling for a fundamental shift in the way the national funding model assessed career interruptions in funding applications:

> I guess one thing to take into consideration is that the NHMRC have a new investigator grant. So you put in your grant application and tick a box to say you are a new investigator ... When the scores come in new investigators get an extra point which is significant when ranked out of seven. It is almost like we need to say let's review all the grants and take all the career interruptions out of the equation and put the point on at the end. In the meantime it is too hard to know who is acknowledging [career interruptions] and who is not. It needs a numerical value on it rather than a subjective [one] (Interview 21).

Thus there was a strong sense that the Baby Boomers needed to hand over power and leadership of the institute to Gen Xs and Gen Ys. There was also a clear message that change was coming. The new science leadership model would include more women in leadership positions, better work-life balance for men and women, and a new funding model that more fairly assessed career interruptions. But, more fundamentally, the younger generations were questioning the model of science excellence itself which they saw as outmoded and highly gendered.

Challenging Old Models of Work and Funding

The Gen Xs and Gen Ys in this research project challenged many of the both stated and unstated assumptions of Baby Boomers about science research careers. One was that male careers were prioritised over the careers of their female partners because they were the breadwinners. This was no longer true. Almost all of the younger men interviewed had partners who had their own careers and who often had more flexible work arrangements. Others, like the following respondent, had wives who earned more money than they did and thus took the financial pressure off them: 'I have a wife who earns lots of money. There is no pressure on me financially. If I was on my own or the main breadwinner, there would be more pressure for me to get a better salary. I'm just lucky' (Interview 15). The corollary of having a wife that earned more was that her career was just as important—or even more important—than his and that the couple needed to negotiate how they combined their careers and other responsibilities.

The AWIS (2012) survey reported that only a third of respondents agreed that they worked for family-friendly institutes. Another third said that their institute did not have a spousal hire policy or that such policies or other types of support were not available because of funding cuts. Not surprisingly, a quarter of respondents (but 46 per cent of those under thirty-six) would consider moving abroad to further their career. One suggestion is that funding bodies need to offer new competitive fellowship schemes specifically for researchers who have to relocate in order to follow a partner (HoC 2014, p. 32), which would provide recognition that many Gen Xs and Gen Ys are now juggling dual careers.

Some respondents in this study challenged the notion that to be a successful scientist you had to be in the top 20 per cent and follow the advice of their supervisors about how to make it to the top. Rather, they questioned old work models and considered that balance and lifestyle were far more important:

> I do believe there is a generational shift which does make it tough, because lab heads say: 'I did it'. My generation were brought up a lot luckier. At the same time, the analogy could be an athlete, you can't stop a person training and wanting to be the best; you can't stop a person wanting to work fourteen-hour days. You can imagine that the top 20 per cent who are getting all that support are very driven. Then there are the 80 per cent who want lifestyle and not getting too stressed and who will fail because there is not enough money around. I think that is changing (Interview 15).

Gen Xs and Gen Ys challenged not only work models, but also the funding model. The clear message was that a funding model which discriminated against researchers who had career interruptions was inequitable and unacceptable. One questioned if the NHMRC assessment of output relative to opportunity was really taken seriously:

> I notice in the application forms they do have these clauses that you can apply that you have had a break due to having a baby but I really wonder how much that works when people are reviewing them. It is definitely a concern for me. That would be a really hard thing to deal with if you were a woman. I would have considered taking time off too if we have another baby if that was not such a problem. But I just don't know how you can manage it. It is tough (Interview 14).

Another considered that the way the NHMRC assessed career interruptions should be improved. Moreover, due to its inflexibility around part-time work on grants, those applying for grants should 'cheat a bit' in declaring that all staff employed on the grant would work fulltime:

> What would be best is to say 'you are pregnant, you have a year off'. They should force every supervisor to allow half-time work. But the NHMRC don't allow this on their grants. The supervisor can extend the grant, but the NHMRC doesn't allow it. So you have to kind of cheat a bit (Interview 18).

Yet another respondent strongly criticised the present funding system and called for a better formula for assessing output relative to opportunity. Furthermore, they asserted that this should be available for both men and women:

> The career disruption issue has been emerging, and certainly is something that goes into the granting reporting process. But I don't think it has been taken as seriously as it should be. I think it applies to men and women. If a man wanted to take time out to be a home dad he should be able to do that, but I am not sure that it is very accepted either. So when people have to take career disruptions there needs to be a better formula. People shouldn't need to show you evidence. Obviously there needs to be some reporting. But it is completely ridiculous at the moment. We need to find a sensible way of looking at that issue (Interview 23).

But until the funding model changed, going part-time at an early stage in a science research career could be risky, as Interviewee 14 above noted. ACOLA (2012, p. 38) has recognised this dilemma for many younger researchers and called for a new flexibility in funding programs to cater for those applicants who want to work part-time or to spread their grant over a longer period.

The following respondent was keen to explore part-time work, but also worried about the impact on career prospects:

> I would like to have a family at some point and when that time came … if the opportunity for part-time work was there I would think very carefully about that. It depends on

> whether I was being employed on an NHMRC or similar scheme in which case a part-time role would probably not be an option. But if I was employed at the Florey I would be interested. Spending time with a child is something I would like to do but at this stage it would destroy my career prospects (Interview 26).

Not only the funding model but the increasing demands of even low-impact journals in accepting articles for publication became all time consuming and made it difficult for men and women to be active parents:

> Yes, there is a generational change. They want to have more balance in their lives and be more involved with their family, but are probably coming up against the expectations that they should be doing it the way it has always been done. This is driven by the funding model. It is also the journals. Even the low-impact journals are expecting so much now. The amount of work demanded of *Cell* and *Nature* are now demanded of low-impact journals. It is so much harder to publish and so much harder to get grants. So where does family fit into that? (Interview 34).

While Gen Xs and Gen Ys were not yet in a position to substantially change their working environment, they were sharp in their analysis of what needed to change. First, science leaders needed to understand that Gen Xs and Gen Ys were juggling dual careers and family. Most had partners who had equally important careers and for some of the men, as described above, their wives earned more than they did. Second, some reported that they did not necessarily want to be in the top 20 per cent; they would rather achieve a balance between work and family. Third, they challenged the current funding model: the evidence required to prove career interruptions, how output relative to opportunity was assessed, and the inflexibility that required PIs to 'cheat' in order to accommodate research staff on their grants that wished to work part-time. Finally, there was criticism of the demands of journals which required often huge additional work to be undertaken on articles that they had provisionally accepted for publication, and the impact this had on family life.

Thinking of More Flexible Work Models

Despite the constraints on Gen Xs and Gen Ys in this study, some were already challenging old models of work and funding in order to more actively engage in parenting and to balance dual careers. While this was often difficult, there was a determination to at least try new models. Take for example the following respondent whose narrative begins by talking of a 'strange' background. What he goes on to describe is a career that is actively challenging the way science leaders and funding bodies mandate how scientists do science:

> I have a slightly strange background. For quite a lot of the last five years I have chosen to work four days a week. In the beginning that was out of choice, but a bit out of the money that my supervisor had for me and it was easier to fund me for four days. I have two kids now. And I have had periods when I worked four days, when my wife went back to work. I haven't noticed any massive difference in output between working .8 and fulltime.
>
> I would rather work at .8 even given less money than fulltime. Just recently my wife has gone back to work after having our second child. I haven't broached the topic with the funding body for my fellowship. But at the moment I work fulltime over four days. I have one day at home looking after the kids, but try to get a few hours work done, mostly in the evening. And then I work four long days for the rest of the week. That is not ideal. I would prefer to work .8 and have one day at home and so that is what I have applied for [in] my next fellowship (Interview 10).

The narrative indicates that this young male scientist is making choices about working hours in consultation with his wife who also has a career. There is a sense that working a compressed week in order to have one day during the week at home with the children is more important than money or career progression. However, he does not believe that working .8 or working full-time over four days per week has in fact had a negative effect on his productivity.

There was much debate throughout the interviews for this project about how part-time one could work before a critical stage was reached and a scientist would fall behind the competitive pack. The previous respondent thought that .8 was manageable. However, another respondent had initially, as a new parent, tried working three days a week but was 'falling behind'.

> I have had a paternity leave. And I am currently still one day a week looking after my child at home, who is three years old. When my wife went back to work when she was three months, I was doing two days a week at home; about half of that I could work when she was asleep but these days that it is impossible. I have had that since she was three month old. It really hasn't impacted my career progression that much, it has probably actually helped. I find I can get as much work done on four as five [days] just by working a little bit here and there on the weekends and when I do come to work I knuckle down and squeeze five days work into four. I think having a child has motivated me a bit more. I don't think it has impacted on my career progression. I found that three days a week was hard and I was falling behind. Four days a week at work is okay. And that has been the majority of the time I have been on paternity leave.
>
> I don't think it is possible to work part-time and have a successful science career. If you want to drive a successful career and become independent and run your own laboratory you have to be full-time. You are just going to get knocked out in the competition. There isn't enough money to go around. Certainly a part-time job in a lab and doing what someone else wants you to do is fine. Part-time career opportunities should be there, with the understanding that if you want an independent career you are shooting yourself in the foot (Interview 15).

This respondent, like the previous one, was working full-time over four long days each week. He saw benefits not only for his parenting, but also his science, in this approach. He was more motivated and could 'knuckle down and squeeze five days into four'. Interestingly, he did not see his preferred way of working as part-time—he was in reality working a compressed week—and argued strongly that going to part-time research was like 'shooting yourself in the foot'.

While these two male scientists are trailblazers at the institute, there are others who are planning to have children and are determined to negotiate working hours so that they can know their children and 'spend time with them'. One respondent had been influenced by these role models:

> And also being married it was thinking about family factors in terms of what would happen when we had kids, who would take time off work ... I look at a couple of men. They have taken a day off. Two days a week in science would be difficult. Yes I would look at it. It would be four days a week. It would be positive (Interview 12).

Another was thinking of working part-time in the future and explained:

> I would definitely like to consider part-time career opportunities. Wanting to establish your own lab requires a huge commitment. I watch the lab heads I work with. They often work fourteen-hour days—essentially the norm to be successful. I am happy to do that at the moment. But I would like to have kids and I would like to spend time with them. The idea is to get the hard work done and then have kids as a male if you want to get to know them. Otherwise as a female, it is hard to take time off even as part-time. You can't get established as part-time. You can only do that as a lab head and you have established people doing the groundwork for you. You can't do it at this stage. You can't tick all the boxes (Interview 3).

A further respondent was also giving some thought as to how to combine work and family in the future, and took the view that it would be easier to do so if they were more senior in the organisation at the time they became a parent: 'I definitely think that family would come first. I am a workaholic at the moment; that is probably because I don't have a family ... But if I managed the group it might be easier to have flexibility and to work from home, etc.' (Interview 4).

Thus, some respondents were planning to have family and thinking about how they would combine parenting with work, and possibly delaying that decision until their careers were further advanced. Their views resonated with the findings of the AWIS survey (2012, p. 4) that found 39 per cent of female respondents and 27 per cent of males had delayed having children in order to pursue their research career.

A third respondent also talked about considering a part-time career in the future but could not find support from their supervisor, again suggesting a clash of generations on this issue:

> At one stage when we have kids I would like to work half-time. Half-time doesn't mean half-time; it would be four days a week or three days. But my supervisor doesn't like it. Because you can't really if you have a big experiment, you can't really do it in two or three days a week. I am not sure if he would allow me to do it. But so far it hasn't come up (Interview 18).

Other discussion about work and family included two respondents who looked forward to working less than full-time when they have children. They were not worried about 'being in the top 5 per cent' or about the response from colleagues to these plans:

> If my wife and I do have children a four-day week would be fantastic and I would do that at the expense of money and potentially at the expense of being a lab head. I can definitely get job satisfaction by not being in the top 5 per cent and having a family life outside that (Interview 26).

> One of the things I would like to do and everyone seems to think it is left of field ... I had a colleague ... who was the stay-at-home dad and went part-time. He wasn't that ambitious, and it didn't affect him too much. It depends how long; if it is only six months or a year it is not going to have a big effect at this stage in my career. Part-time, most definitely. My wife plans to go back and do a clinical PhD which is four years full-time. I would like to think I wouldn't mind going part-time for a while (Interview 29).

Another way of thinking of more flexible work models was to change the model of science careers, so that the trajectory did not need to be ever upwards. Thus other models of career progression, one respondent argued, might help women who took maternity leave:

> Maybe there needs to be a few different steps up and you could be a postdoc for a few years with no added responsibility. And maybe when you want to take that next leap you could. There definitely needs to be better maternity leave arrangements. There needs to be more certainty, maybe rather than writing grants and waiting to see if you get them; there needs to be some extra system set up. I think the leakage is in that early postdoc phase. There really needs to be more flexibility at that phase (Interview 12).

The notion of 'an extra system' suggests that women with interrupted careers due to family responsibilities need a separate funding stream that acknowledges their different career path. The McKeon Review (2013 p. 138) endorsed this view and called for new programs specifically for women with career interruptions due to parenting such as 're-entry' or 'retention' fellowships, or mentorship and support for senior women researchers. It will be interesting to see if the NHMRC adopts this recommendation and creates these new programs.

So quite a massive change is already taking place in how Gen Xs and Gen Ys do science or wish to do science. This is happening despite the leadership of the Baby Boomer generation, who have a more rigid model of working hours and working styles. Some senior scientists are aware of this change process, as indicated below, while others see the change happening at merely an individual rather than a systemic and/or generational level:

> The modern young male is not your monastic scholar. We have male scientists getting NHMRC grants and they are trying to do it as a .6 slot. These young couples, they are juggling dual careers. That has a significant impact. It then flows back on division heads; they have an urgency to make sure their division is successful. There are competitive tensions as to which is the best decision. They don't want to lose good staff. The repercussion is that people are doing five days' work in three days. So I think gender and that period from thirty–forty-five is really, really tricky (Interview 36).

Clearly, the younger generation of men, as well as women, in science saw part-time careers as attractive and as a practical way of managing dual careers. The Royal Academy of Engineers in a submission to a UK government inquiry concurred, and argued that there were advantages for science in men adopting flexible work practices: 'having high-profile men who take advantage of flexible work contracts or who have made it to senior positions via non-traditional routes is really important' (HoC 2014, p. 27). Part of the massive change identified in the current study is around the dual careers of Gen Xs and Gen Ys and their quest to achieve work-life balance.

Achieving Work-life Balance

Achieving work-life balance for the Gen Xs and Gen Ys is a continual process of juggling time and negotiating with their partner about how to make it all work. Some have been able to successfully achieve this balance, while others are fearful of changing their current work arrangements. What follows are four case studies of how Gen Xs and Gen Ys have tackled the challenge of work-life balance. The first is a successful male researcher who talks about the way he and his supervisor had planned for the birth of his first child and how this would impact on his research output; the pressure from peers not to be part-time; and having to grab 'morsels of time' when he could, between experiments and responsibilities as a father. He was therefore empathetic to how much more difficult this must be for a woman with children:

> Me choosing to have kids, it definitely impacted my work productivity; and I am the father. It certainly ... it's probably put me a year behind in my output of publications. When I informed my boss ... that I was going to have kids he said 'okay this is going to put you a year behind, and we are going to plan for it'. That was based on his *own* experience. For a mother it must be much harder. But for me I know how hard it was going from 'I can work whenever I want' to now 'I have to do everything between these two blocks of time'. It is then re-learning how you work, and getting everything done, with timelines on the day. The experiments I do are technically difficult. Invariably you are pulling the greatest results ever at 5 pm and then you have to get your kids at 6 pm. [Do you have to be more strategic?] Absolutely, my competition who don't have kids have a fantastic advantage over me right now. I could afford to basically talk to another scientist for an hour a day, shooting the breeze. For me the biggest impact is how to get my experiments done and then learning to find other morsels of time to complete all the computer-based work you have to do. When you get to the perception of how this would be for a mother; that is clearly going to have an impact on how people perceive your output. And it takes you out of the networking.
>
> Part-time may have been [an option] in the first year of my son's life. I don't think that would be viewed favourably by my peer group. You don't get any slack for that. Five days in four? It is not going to happen. It depends on what experiments you do. Certainly once you get the data you can take the laptop home, but you can't get much work done with the kids around. To tell the truth I need to work more than five days a week (Interview 20).

In the second case study the scientist could see beyond the present dilemma of the respondent above who was time poor, to a time when the children were older and the demands of a science research career would be less, and they could 'level off':

> I would see that over the next five years I will not work any less. But after that I would like to level off. With young children you don't get much sleep and you have to double your output. The bar is very high now. You have to reach that bar and it takes so long to have any independence in Australia ... It is very hard to find what the even playing field is here (Interview 19).

In the third case study the respondent discussed the difficulties of dual careers operating in tandem when both worked in the science research field, and took the view that it was impossible for both to progress at the same pace. One needed to 'step down':

> I hope I didn't sound too pessimistic. If I was to have more children now it would be very challenging, especially when my husband is also a scientist, so that he needs to apply for grants when I am applying for grants. If you have a partner who is prepared to support you it might be different.
> ... At one point you need to decide who has priority over the other person. I kind of step down at this point. He needs also to set up his own lab ... Once he has set up his lab I will set up my lab. If we were both to run high in parallel we would not have children. There is always one that needs to step down. Or you meet later in your careers. If you are both starting one needs to step down (Interview 16).

The way this couple managed the challenge of dual careers in a fiercely competitive science research environment was at this point in time to prioritise one career over another. The respondent does not talk about working part-time. Rather she puts less time and energy into her career to enable her partner to focus on building his career, and presumably takes more responsibility for parenting.

There is a sense that the pendulum might swing in the future and her career might be prioritised.

The final case study highlights another new model of building science research careers. This entails making a joint decision about balancing marriage, careers and family, and it is about two careers rather than one career sacrificed for another career:

> When my wife went back to work we talked about how we could do it. With me working four days from 8 am to 6 pm and having a day at home, that's fine. I get all my work done. Family is okay. We still have weekends to do things. That is one of the positives in what I do.
>
> My wife took a lot of leave. I have certainly seen how good it has been for her to get back to work ... And I think my generation see that; if you can and want, that it is really healthy for you both to have a work environment. In my family life we have a good mix. I work four days, my wife works two and a half days.
>
> Part-time definitely, I think .8 is my ideal. I say that with a little trepidation. I don't have the drive to work fourteen-hour days. That is not really me. The reason why I am a bit hesitant about .8 is that maybe I will get ten years down the track and wonder what would have happened if I had worked full time and had much greater output. But at the moment I think that is unlikely and I value the other things in my life too highly. So yes, I think part-time opportunities are important (Interview 10).

It is clear that negotiation and deft organisation are required by this couple in order for both careers to stay on track. However, they have managed to achieve a balance, expressed succinctly in the comment: 'I get all my work done. Family is okay'.

Summary

This chapter has identified a fundamental shift in the way younger women and men wish to do science which reinforces the findings of other national (HoC 2014) and international (AWIS 2012) research on generational shifts in science research careers. The women and men in this case study reject the traditional notion of scientific excellence and the traditional science construct of masculinity. For them the normative male model of the scientist, the monastic male that is fiercely competitive and has no life apart from research, is in reality a myth. Their views are consistent with González Ramos et al.'s (2013) assertion that excellence in science is 'not a neutral process but a hidden structural mechanism of access and promotion that keeps women in a secondary role in scientific institutions, through hiring practices and a promotion system based on a male design of professional careers'. Some respondents were not fussed if they were not in the top 5 or even 20 per cent of science researchers, or if they did not become lab or division heads. Rather, they were trying to redefine masculinity and femininity within the workplace and what constitutes a successful scientist. Several have managed to work a compressed week, that is, a full-time workload over four days, by working long days and doing some of the work at home in order to balance their career, their partner's career and family demands. However, this was generally not permissible under the terms of their funding and required supervisors to 'cheat' a little. Others could envisage a time later in their career when more flexible work arrangements might be an option. Some of the women, as described in the previous chapter, had chosen to work part-time in order to achieve the balance between work and family and accepted the impact this might have on their productivity. What they asked was that supervisors and colleagues respect their preference for part-time work and continue to support their career development, rather than consider they were not serious about a career. Even those who did not yet have children could envisage that they may wish to work differently when they were parents.

The next chapter examines recent debates in Australia and globally that emphasise the need to keep young scientists in science in order to increase national competitiveness and innovation, and then explores various strategies for keeping younger men and especially younger women in science research.

Chapter 10
Keeping Women in Science Research

National and International Context
This study initially focused on how women and men in a large Australian science research institute build their careers. It then looked at issues of gender in science research. What became clear in analysing the paucity of women in senior positions was that the organisational culture constructed women as single, with the ability to reach the top in science, or as mothers who had made a choice, based on biology rather than gender so the argument goes, to have children. That choice led them to seek careers elsewhere, perhaps combining research with clinical practice, or in industry or academia. Those who held to this belief saw no role for the institution in exploring how women who have children could be encouraged or mentored to stay in science.

So what are the imperatives for keeping more women in science? Recent debates in Australia and globally emphasise the need to keep young women in science in order to increase national competitiveness and innovation. There is a real crisis in the EU, for example, to remain globally competitive. Only 20 per cent of science research is now being conducted in the EU, which is losing out to South East Asia. And while EU scientists produce even more publications than the US, most of this knowledge is being commercialised somewhere else in the world. Significant social reforms and funding stimulus are therefore required in order for the EU to become more competitive and to help commercialise European research (Carvalho 2012). As Vlasak (2012) argues, in the search for more innovation the EU 'cannot afford to waste any more talent'. In the UK it is recognised that the economy needs more skilled scientists and engineers and that 'this need will not be met unless greater efforts are made to recruit and retain women in STEM careers' (HoC 2014, p. 49). Australia faces similar challenges, and in particular needs to increase the percentage of GDP it allocates to research.

Morley (2014) argues there is a gendered research economy. Globally, men have the edge as researchers by a huge ratio of 71 per cent to 29 per cent (UNESCO 2012). Interestingly, the highest proportion of women researchers is in countries that have the lowest research and development expenditure, while the lowest proportion of women researchers is in countries with the highest research and development expenditures, such as Austria (Morley 2014). Women are less likely to be journal editors or cited in high-impact academic journals, and are under-represented on research boards and peer-review structures that allocate funding. Women are also less likely to be principal investigators (*Nature Neuroscience* 2010), are awarded fewer research prizes and receive fewer invitations to be keynote conference speakers (Morley 2014, p. 116).

There is a clear link then between the gendered research economy and funding of research. It has already been suggested throughout this study that the science funding model in Australia needs significant reform in order to recognise and reward women who have had career interruptions. This will require a separate funding stream targeting those who for a relatively short period of their career have been less productive because of career interruptions that result in lower productivity.

In addition, the attitudes of science research leaders need to be challenged. It was evident in chapter 8 on gender and career paths that some supervisors had varying views on their role in mentoring and advising their women postdocs when they were considering having a family or became pregnant. As one respondent explained, it was about 'just changing the mindset of supervisors and people in senior roles that they are open to the idea and that they recognise that women want to have families and how that can be incorporated' (Interview 38). One means of changing this mind-set would be to introduce mandatory diversity and equality training, including unconscious bias training, for all line managers and supervisors of staff as well as for all members of recruitment and promotion panels for STEM jobs in higher education institutions (HoC 2014, p. 50).

This chapter will analyse how new, more flexible models of doing science can be implemented. These include: exploring part-time careers, new models for science careers, women being strategic about when to have children, and effective strategies for transitioning back after maternity leave. Finally, it will present a case study of an effective model for women combining career and family, in which the supervisor played a strong leadership role.

Part-time as a Career Option

Interviewees in this present study were asked if a part-time career was an option for them. The PI was surprised how many respondents, both men and women, were prepared to countenance part-time careers, although not necessarily for themselves. One saw part-time work as a way of staying in a science research career: 'I think part-time career opportunities might be useful, having that other career [clinical practice] and knowing that I won't be doing that full-time for a few years until the kids are off my hands. It would be good to keep a hand in things' (Interview 6). Another saw part-time work as important for women who had been on maternity leave and suggested various strategies to enable women to keep on track while working more flexibly:

> Maternity is a big factor ... Just to be able to work part-time at that time would be helpful. The reason there are less senior women is that women have less experience because they have had to take time off ... and [need] opportunities for part-time research during that time. A lot of your CV is first author—if you job-share you would have joint first author. Not sure how I would share ... If you are doing

more applied science, if there is RA support, you can get them to do the experiments rather than you being in the lab 24/7 (Interview 9).

Others wanted their decision to work part-time both recognised and respected as a valid career option at particular stages of their career. As one woman who had made a decision to work part-time put it, she did not want to 'work like a man' in science:

> I have a great male mentor. When I told him I was pregnant he was genuinely delighted and said 'congratulations' ... There has never been any problem with that. But I can't model my working week on what he does. And the female [role] models work full-time. Also [name deleted] a very successful woman who works fulltime ... I don't want to be like her. The only way I can possibly succeed is to work like a man. I am not going to do that; I won't do that (Interview 35).

It was clear that the leadership to implement more flexible work models needed to come from the supervisors. As the following respondent indicated, some were more open to women working part-time than others. Moreover, there were challenges in keeping research 'ticking over' when a post doc was working part-time, but with good communication and careful planning this could be achieved:

> I think some lab heads are not good like that [providing career mentoring for women who are working part-time]. And one of the reasons why I think this is important is that there are now more women in science than the men, and many of them are very highly trained and they want to work, whereas less men are doing PhDs; they want to go and make mega-bucks in other industries now. I think you have to foster the careers of these really talented women. And to actually enable them to keep publications coming out is a skill ... I think this could easily be managed if you really think about and prepare for how you are going to do it. If you were working for me I would find a project we

would agree upon where you could in your 40 per cent, 50 per cent, 60 per cent very effectively manage it. You don't have to be a key author on a publication. If you have a bit of time out but are still getting publications that is really important. Lab heads have to be made aware of the fact. We don't like years where there are zero publications; even if it is a little paper in a low-impact journal. And by looking at a lot of career interruptions, this needs to be thought about quite carefully (Interview 37).

What is evident from this narrative is that science research institutes cannot afford to lose highly trained women who prefer to work part-time, because male scientists with PhDs are increasingly going into industry where the salaries are more attractive. Moreover, strategies can be implemented to enable women working part-time to keep engaged in research. However, research institutes often do not make it easy for staff that wish to work part-time. The UK Institute for Physics and Engineering in Medicine noted that 'there appears to be a dearth of opportunity for sufficiently flexible working patterns, or real commitment to "family-friendly policies"', while the University of Manchester argued that even where flexibility and part-time work did exist 'there is a perception that career progression is more difficult, as quantity is valued and quality alone is not enough' (HoC 2014, p. 39). In order to keep women in science research, supervisors need to explore flexible models.

New Models

Some respondents were aware of other models of combining family and science research which worked well. Several mentioned the Swedish model of parental leave as providing greater flexibility for both men and women with families (Interviews 7 and 13). Another was aware that women with children had difficulty combining career and family and argued that the institute should foster these women (Interview 14).

A further respondent thought that the institute was well placed to allow its research staff flexibility in their work arrangements, but saw a particular role for individual supervisors in ensuring a contented workforce:

> I think a research institute is in the enviable position of being able to be flexible where its workforce is concerned … So we really should be able to take advantage of using flexible hours and flexible part-time arrangements to suit people's busy family life and busy research work. And that has been something I have been really appreciative of, but it comes down to your supervisor. We should see that we have the potential to be flexible and make people happy and contented workers (Interview 10).

Another respondent also took up the issue of flexibility, as well as the need for on-site child care: 'Things like flexible working times are important. Having child care in this building would have been wonderful. That would make it easier for women with children. More recognition of career interruptions would be good. I can understand why it is a very slow progression for women' (Interview 22).

Some of those interviewed had thought through the process involved in women taking maternity leave and transitioning back. The ACOLA (2012) study identified better schemes to help women transition back after having children as an issue, especially for post-docs. Considerations of respondents in the present study included: who would keep the research ticking over when the woman was absent, having an RA to take over some of the research and how would it be funded, and whether or not a close colleague might take up some of the project. The respondents below suggested that other models should be explored, including students taking on part of the responsibility for the research, and freeing up the hours required in the lab so that a woman returning from maternity leave could spend more time at home focussing on writing papers:

> One of the best things that can be done is to keep a research group ticking over while a woman is on maternity leave, whether that is support for a RA to come in and do the experimental work the woman was doing or have that cover over the gap … if that woman has had a baby and wants to come back and is very driven to do the research. A gap of a year can be difficult. That seems to be the

biggest difference; women often have this enforced time off. I think that needs to be supported. There definitely need to be strategies for women who want to do that. How you fund it; that is tricky. Some money would have to be put towards it. But I think it is a worthwhile thing to do ... If they have a close colleague who can take up some of the project. There must be other models (Interview 10).

We have to be mindful of the fact that there are only so many hours in the day and many competing demands on our time. If you come back part-time, you take work home ... Things like RA help would be good. As far as keeping projects running without the woman being there would definitely help. Also students who are co-supervised by more senior scientists who would take on part of that responsibility would help and mean that the woman doesn't need to be there fulltime. You would reach an outcome sooner and get publications sooner (Interview 22).

Thus, as discussed in this chapter and in chapter 8, there was a realisation that a new model of funding science research was required that targeted both women and men who wished to work part-time.

In summary, over a quarter of the respondents asserted that new models were required to enable women to combine science research and parenting, and that supervisors and the institute should provide leadership on this issue. Moreover, they argued that the science funding model needed to be reformed to acknowledge that both men and women with children may need to work less than full-time.

Being Strategic about When to Have Children

Throughout this study there was a good deal of discussion about when would be the optimal time for a woman in science to have a first child and remain in science. One thought that the second year of a PhD was a good time (Interview 2), but others disagreed. It appeared that there was no good time for a woman to have a child and remain in science research. However, the following respondent mused about what might be an optimal time:

> If you are a woman in science and want a family you have to think about timing. I would have my kids while I was doing my PhD, during the middle of it. Are you going to then drag that family overseas and take that partner with you? One of the reasons why going overseas was easy, was because there was only me. Your next option is to become independent, get the fellowship. And as soon as you have the letter, have the family (Interview 20).

But there was general agreement that the more senior a woman was when having her first child, the easier this juggling act would be because they would have more flexibility in their working hours. The AWIS (2012) study found that a third of female respondents were in fact delaying having children, mostly until they had better salaries. And one woman scientist told a UK inquiry that she had 'an unpleasant choice: risk not having children or risk having to restart my career in my mid-thirties' (HoC 2014, p. 36).

It was evident that women doing PhDs and in the early career phase were often constantly juggling research and children, whereas women who had delayed having children were more likely to quickly consolidate their careers and have children before their fertility declined.

Strategies for Transitioning Back

Critical to building new models for doing science is developing strategies for women transitioning back to work after maternity leave. A whole range of issues needs to be considered. As the following respondent explained, first and foremost, understanding is required; second, support in the lab must be provided so that the woman on maternity leave can direct the research and read the literature at home. Third, funding bodies and the institute need to recognise that this person has had a year off and make adjustments in assessing their research output:

> There has to be something available to plug the gap for women [that] is created by children. The understanding has to be there. Women need to be pro-active in not staying away too long. But you need one or two years of

support. Providing a pair of hands to do the lab work would be good so that they can sit around at home and think about directing the research, and reading the literature. And there needs to be a concession at the other end by the granting bodies and the institute that this person had a year off, and pretend that they are younger than they are. As a reviewer it is hard to remove that long period of maternity leave from your mind. You can only go as far as you can; it is up to the reviewer. I think there are guidelines about how to review these applications. It is just difficult to do (Interview 15).

Others also supported the provision of an RA while a woman was on maternity leave (Interviews 26 and 28), and affordable on-site childcare (Interviews 26, 27, and 29). Another important strategy was supervisors keeping in touch with women on maternity leave (Interviews 26 and 39).

On returning to work, part-time options and flexibility in work hours were suggested as strategies (Interview 29). A further strategy was to provide women with a quiet room where they could breastfeed and on occasion bring in their other children:

> We thought a child friendly room might help. In the old building every postdoc has a room or share of a room. But here [in the institute's new building at Parkville] there is open plan, so it is hard to bring the kid in. So that is why we are trying to set up a family room (Interview 33).

Another important strategy was for the institute to raise funds for a start-up package for women returning from maternity leave. This might include funding to employ an RA and funding for childcare:

> I always thought there should be a separate short grant for a woman coming back from maternity leave. That they will be given a start-up package [and] employ an RA. We have been pushing for that. The institute is trying to get people to donate money and the interest from their capital fund

could finance this. I think they have to raise up to two million to sustain the fund [Subsidised child care as part of that?]. Florey has as part of an agreement that childcare can be salary sacrificed. We have thought about the start-up package [and it] can help with childcare (Interview 33).

As this research project was being conducted, the institute was embarking on a major initiative to keep women in science research. With the generous support of philanthropist Ms Naomi Milgrom, a Women in Science corpus has been established to improve the representation of women in the workforce at the Florey. Ms Milgrom committed $500,000 to the initiative in March 2012. The fund is growing and will provide support for decades to come, encouraging women to advance through the ranks to senior positions. Some initiatives to flow from the funding include financial support to ensure scientific endeavour continues while a scientist is on maternity leave, fellowships for outstanding female scientists, access to advanced leadership training and professional development.

Other research institutions have implemented similar schemes. For example, the Walter and Eliza Hall Institute in Melbourne has technical support available to reduce the effect on productivity caused by women going on maternity leave. This money can be spent on either an RA or a post-doctoral researcher (WEHI 2013). In addition, its Craven and Shearer Award provides childcare support for outstanding female postdoctoral fellows who are aiming to become independent laboratory heads or current female laboratory heads with pre-school-age children. Monash University's Advancing Women in Research Grant program offers funding support and career coaching (Monash 2012) and its Gender Equity Travel Support Grants financially assist female academic/research staff at all levels with primary carer responsibilities to travel (Monash University 2013).

While recognising the importance of the Women in Science fund in improving the representation of women at the Florey, the role of fathers and mothers and the impact of caring responsibilities on their careers need to be acknowledged. Greater flexibility in working conditions may enable them to more effectively balance work/dual careers/family. One initiative, identified by the UK House of

Commons report, was to 'determine and operate appropriate core working hours with flexibility outside those core hours' (HoC 2014, p. 40). The institute, as a result of this research project, has moved to introduce family friendly meeting times as discussed in chapter 12.

Various new models for combining work and family and for women transitioning back to work after maternity leave have been explored above and best-practice initiatives have been noted. But the PI was keen to examine in detail a new model that had been developed effectively within the institute. The chapter therefore concludes with an extraordinary narrative of how one woman and her male supervisor grappled with the challenges that her pregnancy, impending birth and absence from the office presented.

New Models: A Case Study

This case study comprehensively brings together elements of the new models for combining work and family and transitioning back that have been canvassed so far in this and the previous chapter. The following headings have been added: networks, developing a good model, communication, and impact on productivity. Each of these headings relate to the respondent's narrative of informing her supervisor that she was pregnant, opportunities to continue networking while she was pregnant, planning how the research would continue when she was on maternity leave, developing strategies for keeping in touch, and then planning the transition back to work:

> I took six months off from the end of 2010 for maternity leave. And I think because of the way the head of division let me structure my maternity leave and my return back to work, it hasn't had too much of an impact on my career at all. So I did some paperwork and kept my finger on the pulse while I was on maternity leave. And then he has made it very easy for me to come back to work on a part-time basis. I came back initially three days a week, and now I am working four days a week. But because of the nature of my job which has a lot of computer analysis … I can do a fair bit of my work from home as long as I have internet and phone access. And that was how it worked while I was on maternity leave. I was able to continue supervision of

> my RA, and she was doing the hands-on work. And then we would have a fortnightly teleconference and I would be in constant contact with her by email during maternity leave. It could have been very disruptive but I actually still did manage to do a lot of work during that time and progress things during maternity leave (Interview 39).

The PI then probed further, exploring if this represented a good model for maternity leave and transitioning back, and what were the important components. The first component was that the supervisor did not deny this woman important opportunities to build international networks because she was pregnant:

Networks

> In the year I told my lab head I was pregnant there were actually two international conferences that were beneficial at that time for the work I was doing. And there was no question. I was sent to both and covered for both of them. We spoke about my commitment to the research and my commitment to continuing my project. There was no hesitation about sending me to that second conference even though I was five months pregnant. We knew that the work would continue and that I was not just going to go to the conference, going on maternity leave and that nothing would happen for a while (Interview 39).

Note the difference in approach here compared to the supervisor in the previous chapter who spoke about a female member of his team missing out on opportunities to go to conferences because she was pregnant. In contrast, here there is a clear contract developed where the woman commits to the project and the supervisor supports her going overseas to two conferences.

The respondent described in more detail how this new model was both planned and negotiated:

Developing a Good Model

> Yes, as soon as I had let him know that I was pregnant we right from that very early time we already started discussing how it would work prior to me leaving, and making sure that all of the tools were in place for my RA to be able to continue doing the work without me physically here all of the time. And once we made sure that was all under control we also set up a model where I would be calling up, being in contact. And I had already agreed that I would come in for any meetings that were required face-to-face during maternity leave. That was all really well discussed before I went on maternity leave. And we had already discussed as well how many hours I would be putting in when I came back. That makes all the difference if your division leader is a very sympathetic person to the kind of time you need to put in at home versus at work; it can work really, really well. He is also the type of person who says if I need to leave to go home and look after my little one that is not a problem. I can leave early, come in later, because we have this model in place, because he knows I will do the work from home if required (Interview 39).

Critical to this model is flexibility, which has been identified earlier in this study as one of the key benefits scientists perceive of working at the institute. Hand in hand with the supervisor's flexibility was the knowledge that the work would be done, and sometimes from home.

Communication

The supervisor and the respondent had an extraordinarily highly developed level of communication. It was critical that the communication commenced as soon as the supervisor knew she was pregnant. The supervisor, the RA who would take over some of the work, and the pregnant woman planned for how the project would proceed once she went on maternity leave:

> It was definitely about communication. For the six months before I went on maternity leave a lot of that time with our

monthly meetings was spent making sure that each aspect of what I was working on was under control to a point where my RA could continue on with instruction remotely. It was getting my RA up to speed with all the techniques she would need to do, and all of the projects, what was involved, who the contact people were if I was not available. Just having everything in place by the time I finished. And probably a little bit of flexibility on my side as well. So instead of saying, 'no I am not going to be in contact for six months', I was quite happy to come in for meetings and be in contact with my RA. We had a work retreat in December and I was technically on maternity leave, but I went to the retreat (Interview 39).

Impact on Productivity
Finally, the PI asked the respondent what impact she considered that the period of maternity leave had on her research productivity. She estimated that it set her back by about three months:

> I still managed to get a paper published during that time. So that was a paper in collaboration with other people in the division and a PhD paper I co-supervise. I wouldn't say it has taken the equivalent of six months out of my work, maybe half that. During that time I still put in a fellowship application and read grant applications that colleagues were putting in. I was still contributing even though I was not physically here.
>
> Part-time is dependent on the type of research project you have. I am lucky that I can do a lot of my analysis from home on a computer supplied by the institute. Part-time is definitely very useful.
>
> No, in my experience gender has not been a factor ... I was still given the opportunity to contribute, so it didn't really delay me. If you didn't have that model and you were having a few children it would definitely impact on your career progression (Interview 39).

It can be concluded therefore, from this best practice model, that maternity leave does not necessarily impact significantly on research productivity. Three months of lost productivity, or perhaps six or more months for two children, from a forty year career is a small amount of time. Moreover, this model contradicts the view that a career break may mean 'you are never going to catch up' (Interview 40) or puts 'lead in the saddle' (Interview 32).

If the institute wishes to keep women in science research it must challenge the pervasive stereotypes within the organisation that women who wish to have children are either not serious about a career in science research or are not deserving of sponsorship, mentoring and support while they are on maternity leave and particularly when they are transitioning back part-time. This case study demonstrates that with good communication and political will, supervisors and women who go on maternity leave can work together to ensure that the woman scientist remains supported, engaged with her research and maintains a high level of research output. The institute should establish a dialogue with its research leaders about how best it can support women research scientists and how it can ensure that they maintain effective career progression across the life-course.

CHAPTER 11
New Ways of Doing Science

Overview of the Project
This study set out to examine the hypothesis that women in science research institutes experience greater challenges than men in building effective career paths. Its aim was to focus on the Florey Institute of Neuroscience and Mental Health at the University of Melbourne, Parkville, and the Austin Hospital, Heidelberg, as a case study of how to improve institutional practices that can support career progression, particularly for women in science. It set out to identify current barriers to career progression, including gender and equity and disparity of outcomes, and recommend strategies that can be implemented to address these barriers.

The objectives of the research were firstly to examine the broader organisational culture in which women and men build research careers and its impact on career progression, especially for those who have family responsibilities or who wish to have children. It then focused on external and internal barriers for those in clinical science compared to basic science and strategies that the institute can implement to address these barriers. It was particularly interested in investigating the impact of the organisational culture on women undertaking science PhDs and as early career researchers.

Next, the project examined the gender inequity of women generally not being as well promoted as men in the field of science research, especially on its Parkville campus where few women are research fellows. It also examined how family responsibilities significantly hinder the productivity of scientists— more often women—and their career progression through, for example, fellowship schemes. In addition it explored how the organisational culture could be made more compatible with the needs of women and men in the institute. Finally it analysed effective initiatives to improve career paths for women in science to optimise career outcomes.

In exploring these objectives the study asked forty male and female research scientists at all levels in the institute to discuss their careers. They were asked what gave them most job satisfaction, what internal and external factors impacted on career progression and how important was financial reward. Other questions looked at career progression: the impact of career disruption and the challenges for building careers for women and men who have family responsibilities or who wish to have children. Further questions examined the perceived and actual barriers to promotion and how measures of success could be redefined.

One of the key issues for building science research careers identified in the literature is the role networks play in women's and men's promotion in science research institutes (Sabatier et al. 2006; Wroblewski 2010; Sagebiel et al. 2011) and this study explored whether or not women have the same opportunity as male colleagues to be introduced to them. As part of this networking, the mobility of scientists to take advantage of international networking opportunities is critical and interviewees were asked if this was more difficult for women in science. Mentoring can also play a crucial role in building networks and interviewees were questioned about their experience of mentoring and being mentored, and asked to assess its relevance to career progression.

In the latter part of each interview the PI tried to ascertain if women have a different experience in building science research careers and what were the characteristics of the organisational culture in science research institutes. As well, interviewees were asked about strategies required to ensure that women undertaking science PhDs and in the early career research phase had the same mentoring and support as their male colleagues, given this is the time when they are likely to start a family (Caprile 2012; Cory 2011), and what would be effective initiatives to improve the position of women in science.

Key findings

One of the strongest findings of this research was that scientists are passionate and single-minded about doing science. It is hard to imagine another profession other than a religious vocation where people demonstrate such extraordinary commitment to what they do. Unlike the findings of the 2010 *Nature* survey discussed in chapter 5, the research scientists in this study did not identify 'guidance received from superiors or co-workers' as the biggest influence overall on satisfaction levels, although they did regard providing mentoring and being mentored as crucial to building careers. Nor did the respondents place much importance on financial reward which those in the *Nature* survey identified as the second highest influence on satisfaction levels. Rather, it was the excitement of being at new frontiers of science, being on the edge of new discoveries and being in an important scientific field that drove their passion and gave them most satisfaction. This single-mindedness did at times have a religious edge to it and has led this study to coin the phrase monastic male as the model of the ideal type of scientist; this certainly seems to be the model favoured by funding bodies in Australia. Moreover, several more senior researchers argued that the single-mindedness required for science research was absolutely essential for effective career progression.

But the corollary of the ideal scientist as the monastic male is that scientists do not have any distractions from doing science and certainly do not have families. Therefore, according to this thinking, scientists exist in a rarefied environment like a religious order with little contact with the outside world. But the model is seriously flawed. Men and women in science research do have families, partners, children and wide circles of friends who are not scientists. Moreover, younger men and women are often in relationships where both parties are trying to push forward with their careers at the same time. This model of the ideal scientist as male is reinforced by the media (GEA 2010). It therefore produces particular difficulties for women in science (Moir 2006) and positions them on the periphery of scientific endeavour. This is discussed further later in the chapter.

A second important finding was that an understanding of how to build science careers largely depended on at what point scientists were in their career trajectory. Most of the PhD students interviewed had not given a great deal of thought to their future career development. In contrast, early career researchers demonstrated more awareness of what a science research career entailed and some were already looking at the next career move. At SRO level respondents were on track to becoming independent researchers. All mentioned strong mentoring and support from supervisors, as well as hard work and long hours, as key to their success and to their remaining in science research. While research fellows had mostly reached the pinnacle of their career, uncertain funding remained a concern. There was evidence of generational differences here. Those research fellows now in their fifties appeared to have reached what Riordan (2011) calls career plateauing, and were content with what they had achieved, whereas younger research fellows felt the heavy weight of their responsibilities and the long hours involved in doing all that the role entailed. Moreover, more senior women saw future career development as problematic, citing lack of transparency in processes, especially around promotion, and lack of support for career development, reflecting the findings of Bagilhole and White (2011) in relation to women in university management, a theme that is explored later in the chapter.

Early career researchers might benefit from more frequent and structured mentoring about options for career paths. While the mentoring that the institute provides is generally good, more formal career development sessions with their supervisors as well as occasional career development seminars organised by the institute might be beneficial. These younger researchers are tomorrow's research leaders, and Browning et al. (2012) assert that one factor impacting on career paths of research leaders is an institutionally supportive research culture. The UK House of Commons report (2014) also emphasised that universities and research institutes needed to take responsibility for improving STEM careers for all researchers.

A third critical finding was that networks, mobility and mentoring are fundamental to building science research careers.

Moreover, all three are linked. Building research networks nationally, but more importantly internationally, are essential to becoming an independent researcher. There was evidence in this study, confirming other research (Benschop & Brouns 2003; Sagebiel et al. 2011), that some women had more difficulty accessing vital international networks. This often resulted from male dominance of particular fields of science and male patterns of social behaviour that can exclude women, and reflects Ely and Meyerson's (2000) observation about the difficulty of disrupting the imbalance in social relations between men and women.

In order to build international networks early career science researchers need to be mobile. Australia's geographic isolation becomes a challenge. In Europe, for example, researchers generally need to take only a short flight to participate in a conference or visit a lab. But for Australian science researchers accessing these networks entails a costly twenty-four-hour flight and resulting jetlag. Most young researchers need to go where the top science in their field is carried out—either in the US or in Europe. Given the large distances involved, it is not possible to take up a postdoc in a lab in these countries unless the researcher relocates for several years. In this current study several models of mobility were presented. More senior science researchers had spent between two and six years overseas in some of the most prestigious labs, developed a number of important international collaborations and then had an accelerated career path on their return to Australia. There were also younger scientists who had come to the institute from Europe as part of their career development.

However, it was clear that these models of mobility are, as Ackers (2010) asserts, highly gendered because they presuppose that the early career researcher is not in a relationship, does not have a partner who also has a career, and has no children. What became obvious in this research was that these models do not necessarily suit the younger generation of men, as well as women, in science research, as discussed later in this chapter. Other models of mobility were more flexible and included short assignments—three to six months—in labs overseas developing collaborations or learning new techniques; or attaching a visit to a lab to presenting at an overseas conference. It has been argued that research funding bodies should remove from fellowship conditions any requirement for researchers to move institute or country and instead provide funding for shorter visits to other institutes for collaborative purposes (HoC 2014, p. 51).

This study found that women generally had fewer opportunities to take up postdocs overseas because the point at which this became important to career progression often coincided with the time they were in relationships and wished to have children. These findings confirm Caprile's (2012) assertion about how the model of the ideal scientist or what she called the 'myth of total availability in the scientific lifestyle' penalises parents 'but also women as potential mothers'. She further asserted that: 'Many young women end up believing that science is incompatible with family life and feeling that they have to leave academia if they wish to have a family'. She concluded that 'family related mobility and time constraints may act as a filter in early selection procedures' (Caprile 2012, p.18). There is an implied link here between the tendency of science leadership to regard work-life balance as only an issue for women, and mobility as one of the significant 'filtering' or gate-keeping processes that encourage women to leave science research.

So the focus on mobility in this present study is important. It may be a gendered means of filtering out young scientists, especially young women, as Caprile (2012) implies. Certainly several women said that while mobility was critical, it was not possible for them to uproot their family and move to an overseas lab. And at this point some had begun looking at alternative careers in science. Others decided to combine research with clinical practice and yet others were looking for career opportunities in communicating science, suggesting that institutions need to define, support and equally value new career pathways in science research (Barakat 2014). However, this study found that several women did manage, with partners and families, to do an overseas postdoc, and as a result they had benefited in their career progression, through what Zippel (2012) calls the mobility bonus.

Mentoring was considered essential at all levels. Most PhD students and early career researchers in the institute believed that they were provided with good mentoring. However, some were confused about the difference between mentoring provided by their direct supervisors and mentoring provided by someone outside their direct work area, despite the different role of each type of mentoring (*Nature* 2011). The separate yet complementary roles of the supervisor and mentor need to be emphasised to younger researchers. This confusion about the roles of supervisor/mentor may mean that young researchers are missing out on important advice about career development. Some more senior women researchers argued that they had not received the mentoring required for further career development, but had the resources to find their own mentors or sponsors and sometimes searched for them outside the organisation. But the limitations of sponsorship at this level need to be acknowledged; it is unlikely to transform the culture of the organisation in which they are developing their careers (van den Brink 2014).

It is clear that the institute does mentoring well. However, it might wish to consider providing resources to the various internal groups that organise mentoring to ensure a basic level of training for both mentors and mentees about how to build a successful mentoring relationship, as well as advice about matching mentors and mentees. The institute may also benefit from evaluating the effectiveness of its mentoring programs.

A fourth finding of this research was that gender was often a factor in building career paths in science research, as discussed in chapter 8. The absence of women in senior science research is not just an issue for the institute in this study; it is universal in science research organisations. As Caprile (2012, p.16) asserts: 'women remain more severely underrepresented among research than among other highly qualified professionals'. This was clear in the present study that found women were more likely to be in senior positions at the Austin campus, which combines research and clinical practice, than at the Parkville campus, which focuses on basic research. The severe under-representation of senior women at Parkville was considered an embarrassment to the institute. Is this another example of filtering based on gender that occurs for women in science research institutes? Gender was played out in other ways as well; for example, some respondents had a tendency to blame the women themselves. Another strong narrative was around gender and choice. Several senior researchers believed that women had a choice to be a research scientist or to be a mother, but could not do both. This either/or paradigm needs to be abandoned and the concept of excellence and scientific merit needs to move beyond the patriarchal logic about choice (Benavente & González Ramos 2014). It is evident in the current study that younger female and male researchers rejected this simplistic dichotomy and argued that the discussion needed to change from choices and work-life balance as a personal issue to questioning scientific practice and the assumption that a successful scientist is a monastic male. These findings have been echoed in Cidlinská and Linková's (2013) research that demonstrated women and men were equally critical of the dominant ideas of the scientific profession as a vocation, the normalised academic path and the associated normalised ideas of success in science.

Both men and women saw the under-representation of women in senior positions as relating to an organisational culture that supports homosociability and homophilous networks (Grummell et al. 2009). This can be highly gendered, as men in positions of authority tend to select those with similar characteristics and values to their own (Witz & Savage 1992) and thus can exclude women from senior positions. Some women questioned the transparency of the institute, particularly in relation to promotions, and did not consider women were always given the encouragement they needed to move to more senior roles. The impact of a culture that was extremely competitive and which could at times appear to be discriminatory to women was to produce a sense of fatigue and discouragement for some women trying to consolidate their careers.

The study found that while supervisors were aware of the adverse impact of interrupted careers for women—and men—on career progression, they considered that the funding model and extreme competitiveness it created limited their effectiveness in helping women plan for and return from maternity leave. Moreover, some argued strongly that the institute needed to be creative in developing strategies to retain women scientists, because they were more likely than men to remain in science research. The institute therefore could usefully engage in discussion about these differing perceptions of the organisational culture to ensure more transparency in its policies and procedures, particularly in relation to promotion.

Perhaps the most significant finding was the huge generational change that is underway in science research. While a recently published meta-analysis of gender and science research (Caprile

2012) did not identify generational change as an important issue, the AWIS (2012) survey certainly did. It is clear in the present study that younger research scientists wish to debunk the myth of the monastic male scientist as outdated and incompatible with how they wish to do science. The myth was initially based on the premise that the successful male scientist had no other responsibilities. Rather, doing science was akin to a vocation. The myth was then adjusted to the male who had a supportive partner who sacrificed her career for his career. That female partner either did not work or worked in jobs that enabled them to take most of the responsibility for the home and the family, and thus allowed the primary breadwinner to work fourteen-hour days, travel overseas and have little responsibility for family life. But the traditional model of the science researcher is now an anachronism. Instead, as Moir (2006, p.8) argues: 'in the discourse of scientific practice, we need to re-calibrate the work-life balance scales by first recognising the ways in which the normative male model of scientific work practice is held in place ... It is not that work-life balance is the problematic issue for women in science but rather the very ground upon which the scales stand: the scientist as "he" that is taken as the unquestioned foundation'. Younger men and women simply reject this traditional science construct of masculinity.

In its place, much more fluid roles for men and women in science research are emerging. And this change is a generational one between the Baby Boomers who are still in leadership positions at the institute, and Gen Xs and Gen Ys. In this research there were younger men who wished to have more flexibility in their working lives but were too afraid to explore new models, for fear of jeopardising their careers. But others had taken the risk of pushing out the boundaries of the established model of doing research. Several were working full-time jobs over a four-day week, to free up one day for them to be at home caring for their children. Others were working a four-day week. The difficulty here was that their salary came from grants which presumed that all research staff on the grant would work full-time. While the McKeon Review (2013) recommended the introduction of part-time grants and also the introduction of five-year grants to replace the current three-year grants, the government's response is not clear.

And why were younger men prepared to push the boundaries? They did so mostly because they wished to be active parents and/or because they were in relationships where they and their partners were juggling dual careers. In fact, several had partners who earned more than they did. So the variation of the monastic male who had a supportive, often stay-at-home wife, no longer applies. This present generation of men considered that their female partner's career is just as important as theirs. Moreover, they were empathetic to female colleagues who were trying to juggle career and family. The incongruity between an outdated funding model and supervisors trying to manage staff within this framework on the one hand, and the needs of younger research scientists on the other, leads to increased stress and work-life conflict, and confirms the findings of another Australian study (Herbert et al. 2014). AWIS (2012) also found that most science researchers experience stress and that a conflict of work and life demands was a weekly occurrence for more than half those surveyed. Moreover, those in Medicine and Allied Health were most likely to report this conflict at least weekly (61 per cent).

Male science researchers who did not currently have children indicated that when they became fathers they would like to work fewer hours. This suggests that, for Gen Xs and Gen Ys in particular, more traditional notions of work and family do not apply (see Riordan 2011). In addition, some younger scientists questioned the intense competition in science research careers and were looking for more flexible work models. They believed it was possible to have a successful career without being monastic and being in the top 5 or even 20 per cent in their field. Moreover, they valued a collaborative work environment and could point to the Austin campus where several senior women worked in a collaborative style and were forging new leadership models.

Younger women were just as vocal as their male colleagues in calling for change. Some were struggling to balance all the parts of their lives. Those who had delayed having children until they were more senior in the institute had more options for determining how to keep on track with research. Other women had had children when they were postdocs and chose to come back part-time so that they could achieve a balance between work and family that suited their circumstances. They insisted that their colleagues and the institute respect their decision at this stage of their career to work part-time and not to side-line them. These findings reflect the European Commission's (2005b) call for human resources development to consider how men and women in science can reconcile their work and family responsibilities and have fulfilling careers, and underline how family-friendly HR policies are often at odds with the expectations of organisations that employees either focus on their careers or are less career oriented and wish to participate in family life (von Alemann & Beaufays 2014).

The previous chapter on new models of doing science research demonstrated that a broad spectrum of research scientists across the institute was committed to exploring how to keep men and especially women who have children in science research. Various new models for combining work and family and for women transitioning back to work after maternity leave were examined, including funding for an RA while women are on maternity leave, returning from maternity leave part-time and flexibility to work at home when required. Another was for the institute to raise funds for a start-up package for women returning from maternity leave. The chapter ended with an impressive model developed by one woman scientist and her supervisor. The institute may wish to consider how it supports supervisors when a member of their team becomes pregnant, and it may wish to promote best practice models of a supervisor and a pregnant woman negotiating how she will continue with the research before and after the birth of her child and when transitioning back to work.

Imperatives for Keeping Women in Science

As outlined in chapter 1, it is important to keep women in science research because Australia cannot otherwise maintain an internationally competitive economy, which begins with an internationally competitive innovation system and internationally competitive universities (*Powering Ideas* 2009). Furthermore, this 'failure to make full use of the available qualified human potential of women scientists' is detrimental not only to economies and societies but to individual research institutions (Jochimsen & Muhlenbruch 2009, p. 12). As the NHMRC acknowledges, 'we cannot afford to waste any Australian talent' (Billiards 2014). The nation therefore cannot afford to invest in educating women to the level of science PhD graduates and then see them exit science in much higher proportions than their male counterparts. What Cory (2011) calls the 'precipitous drop off' occurs when women in their early thirties are trying to juggle family and career. A further reason for ensuring that young women remain in science research is that this institute—and science research institutes more generally in Australia (Bell 2009)—has difficulty attracting men to undertake PhDs. As indicated in chapter 4, between 2007 and 2011 women represented between 65 and 69 per cent of PhD students.

To reverse this trend institutions and funding bodies must stop considering work-life balance as a personal concern and instead see the issue as a question of choices that men and women make with consequences for their careers. Moreover research leaders and governments need to reconstruct definitions of a successful scientist and acknowledge that younger men and women reject the traditional science construct of masculinity in science careers. Furthermore, they need to look more closely at the careers of women science researchers across the life course, particularly during what Caprile (2012) calls the rush hour, when career and family collide. What has been demonstrated in this study is that if women are supported by their supervisors and the institute prior, during and after maternity leave, the loss of their research productivity can be small when considered over their entire career span. Such flexibility in viewing science careers across the life course is, as Ceci and Williams (2011) assert, essential to keeping women engaged and having a sense of being valued as science researchers. In addition, research leaders and governments need to acknowledge both direct and indirect discrimination which women with families—and those who do not have families—experience, and examine the organisational culture that perpetuates this discrimination. For example, in molecular and cell biology why do women represent almost 30 per cent of authorships but only 16.5 per cent of last authors (Wilson 2012)? As the UK House of Commons Report acknowledged, the lack of gender diversity in STEM results from 'perceptions and biases combined with the impracticalities of combining a career with family' (HoC 2014, p. 3).

Smart research institutes see the value of investing in the career development of all their research staff, including those who prefer to work less than full-time at particular phases in their career. It is vitally important for these institutes to respect and support staff that wish to work part-time and keep them engaged as valued members of the team. A more restrictive model that labels those working part-time as less committed does not help the confidence or motivation of these staff. However, research institutes in Australia are hampered in investing in their staff by a funding model that is out of step with how young, passionately committed scientists wish to do science. This funding model does not encourage flexible work options for research scientists at any level in science labs. Hopefully, the recommendations of the McKeon review (2013) that address some of these issues will be implemented by the Australian government.

Outcome

This was the first occasion in which the institute had undertaken a comprehensive study of staff to investigate such issues as job satisfaction, career support, career aspirations, flexible career options and gender issues in relation to careers and the organisational culture of the workplace. In jointly developing this project the PI and the EqIS committee had anticipated a number of key outcomes for staff at the institute. These were: identifying where women's and men's career paths differ, why this occurs, and the tools/strategies that can be used to help address the situation; recommending strategies to enhance career progression for women scientists within the institute; presenting a final report with recommendations to the institute's Executive; and developing a best practice model of effective career paths for research institutes.

Chapters 7, 8 and 9 have analysed differences in career paths of women and men, and included discussion of networking, mobility, and mentoring; gender being a factor in building career paths in science research; and the huge generational change that is underway in science research. Chapter 10 outlined strategies which, if implemented, could enhance career progression for women research scientists in the institute, and also presented several best practice models of effective career paths for women and men who wish to combine work and family responsibilities.

Thus most of the anticipated outcomes of the research have been achieved. The project provided rich data on careers and career aspirations. It was hoped that the project's innovation would be in the institute analysing its values and strategic direction, this analysis then becoming a catalyst for transforming the organisational culture in order to optimise career outcomes for all research staff. As well, recommendations were made to the institute's Executive and are discussed below.

Recommendations and Implementation

This study was originally presented as a draft report to the institute's EqIS committee, which provided feedback, especially on the recommendations. These recommendations were grouped under four headings: career development and mentoring, the under-representation of senior women, generational change in science research and support for flexible career options, which are outlined below. Targets were set for some recommendations in an effort to facilitate change in the organisational culture. It is envisaged that these recommendations could also be of value to other research institutes.

It was recommended that the institute consider strategies to provide opportunities for early career researchers to build effective career paths by:
- publicising to current and prospective employees that it provides a comprehensive mentoring program;
- communicating to all its existing staff that mentoring is available to both women and men at all levels and encouraging them to take advantage of the mentoring offered;
- assigning a mentor to new research staff members within a month of their appointment;
- advising supervisors of the importance of staff having a mentor that is external to the lab in which they work;
- ensuring that all its mentoring programs are adequately resourced to provide a formal introduction to mentoring for both mentees and mentors, and that there is rigour and transparency in the matching process;
- providing all early career researchers with a formal career development program that provides information about various career options and mentoring to support career development;
- undertaking every two years an evaluation of its mentoring programs.

It was also recommended that the institute as a matter of urgency address the negative impact on the organisation of the severe under-representation of women as senior researchers on the Parkville campus by:
- appointing new female committee chairs (to committees such as the Ethics committee, Promotions committee, and Occupational Health and Safety committee) starting with a target of one new chair in 2013, and an additional chair each year until at least three committees are chaired/co-chaired by women;
- appointing new female laboratory heads, with a target of one new female laboratory head in 2013 and an additional one each year until 50 per cent representation is achieved;
- disseminating more information about, and ensuring greater transparency in, its promotion policy (for example, ensuring that guidelines for applications to the Scientific Promotions committee and the outcome of the committee's deliberations are placed on the intranet, curriculum vitae displayed, and track record assessed relative to opportunity—for example, number of citations divided by the years active);
- organising annually an organisational climate survey about the differing perceptions of the organisational culture, including direct and indirect discrimination in the workplace identified in this report, to ensure more transparency in its policies and procedures, especially in relation to promotions.

Due to the urgency of this matter the institute felt it necessary to begin actions to address the negative impact that the under-representation of women senior researchers was having on the Parkville campus. Since receiving the report it has developed two new policies regarding gender equality, especially with regard to committee representation. Committees are now working towards developing their terms of reference and working towards achieving an equal gender split (where possible). Working groups have also been established with female chairs. The institute has also now clearly established the criteria of what constitutes a lab head with two new female lab heads being appointed in the 2013–2014 period.

A further recommendation was that the institute identify issues relating to the generational gap in its workforce; recognise the significant generational change occurring in science research globally, and develop and implement a smart work strategy that invests in career development for all research staff and includes a suite of policies that address the requirements of younger workers, including: dual careers, spousal hire, part-time work and flexible work modes.

Finally, it was recommended that the institute and supervisors be encouraged to explore strategies to keep women in science research rather than to consider that their choices are a research career *or* motherhood. These strategies should include: continuing to lobby the NHMRC to introduce a separate funding stream for research scientists who are unable to work full-time and/or have had career interruptions, similar to the very successful Daphne Jackson Fellowship in the UK (2013); setting aside a percentage of grant money for transitioning back to work packages for women and fellowships for women who have had career disruption; exploring the option of internal part-time or joint grants where two researchers can apply, based on a shared research project; supporting supervisors to explore flexible work models to keep women in science research; and publicising best practice models of supervisors supporting female research staff during maternity or other leave and transitioning back to work.

The revised report was then endorsed by the EqIS committee which resolved—consistent with the anticipated outcomes of the project—that it should be forwarded to the faculty of the institute for consideration. The PI made a presentation of the results and recommendations to faculty in March 2013. The faculty then decided to set up a working group to examine the recommendations and prepare a report for the institute's Executive. The institute's response to the report is outlined in chapter 12.

Conclusion

This research investigated the hypothesis that women in science research institutes experience greater challenges than men in building effective career paths. It can be concluded that the hypothesis is proven. Women, especially those with children, do experience greater challenges in building their careers. These challenges often stem from their wish as early career researchers to combine their career with starting a family. Issues around transitioning back to the workforce after maternity leave also present a challenge, as women are looking to return initially part-time and have flexibility in working hours and the ability on occasion to work out of home. They are also asking that their decision to work part-time be respected by management and their colleagues. But some of the challenges result from direct and indirect discrimination that both women with and without children can experience, especially from an organisational culture that is narrow and masculine and positions women as outsiders in science research.

Finally, younger women and men reject the traditional science construct of masculinity; that is, they reject the notion that a successful scientist is a single male whose focus on science is akin to a religious vocation. They also reject that part of being a successful scientist is to be fiercely competitive and in the top 5, 10 or 20 per cent. Instead they are seeking new models for doing science that support dual careers, work flexibility and work-life balance. The role of supervisors in this transition to new models of science research is critical. The institute has recognised the challenge of keeping women in science and has established a fund to improve their representation in its workforce.

A number of recommendations were made to focus on how the institute might more effectively keep women in science. Some of these have now been implemented.

Afterword

We were delighted when Dr Kate White chose the Florey to conduct her research which is beautifully presented in this book. Interestingly, the release of the report has had a galvanising effect on the whole institute. This has helped us implement change more rapidly than we would have been able to do unaided with our existing program in the area of gender equity. Given that the report identified key areas of concern and specific recommendations were made, these were reviewed and endorsed by our Executive as well as the board of the Florey which gave us considerable authority to implement our plans, some of which are mentioned below. The framework within which change is occurring is being established through the Equality in Science Committee (EqIS).

Change towards a more gender-balanced working environment recognises that diversity leads to great discoveries. There is a wealth of evidence in support of interdisciplinary teams. These teams come up with the best ideas and solutions because of their range of approaches and thinking styles. Thus, the Florey recognises we must incorporate a rich and diverse network of individuals so we are best placed to be at the forefront of improving life through brain research.

However, there are many obstacles faced by researchers that often limit them reaching their full potential. The issue of women in science, especially at senior levels, is a prime example. While the number of women and men undertaking PhDs in medical science has been gender balanced for the past decade, the numbers of women successfully climbing through the ranks to the top of the academic tree are small.

Why are we failing to retain so many talented women? The reasons are complex and multifaceted. Unconscious bias, normalised discriminatory behaviours, absence of role models and mentoring, and pressures of balancing job insecurity with primary carer responsibilities impact on the success of women in science. Some of our achievements so far at the Florey are:

Increasing Visibility of Women in Science
We are extremely lucky to have many high achievers in our ranks and increasing the visibility of women in science is a key objective of ours. With the generous support of a committed group of supporters we have raised more than $900,000 towards an endowment to support female leaders in our institute. We are also highlighting the achievements of our women in science through profiles on social media and our websites.

Mentoring
EqIS recognises that diversity in mentoring is key to outcomes. Our initial pilot program which established mentor and mentee pairs has described mutual benefits as a result of participating in our program. We have now established an active working group that is developing a larger scale program with more formal mentoring objectives.

Transparency in Career Progression and Decision Making
To improve transparency within our institute, we have developed position descriptions that clearly define leadership positions. We have committed to achieving gender equity on all decision-making bodies, and hope to achieve at least a 30:70 representation on all institute committees within the next year. Our target of 50 per cent has been supported by Harold Mitchell AC, chair of the Florey board, who has committed to lead this change, starting with the gender representation on the board. We have also established clear definitions of, for example, what constitutes a lab head and have developed

a policy development committee who have already had three policies approved by Executive that are currently being implemented within the institute that target gender equality, flexibility at work and career progression.

Generational Shift
To reflect a generational shift in the way younger scientists view family responsibilities, we have increased flexibility at work by introducing family-friendly meeting times, flexible working hours and part-time positions, and are creating family rooms for working parents. We will continue to support staff in managing their careers and families.

Beyond
We are also addressing broader gender equity issues by joining forces with three of the largest medical research institutes in Australia; the Walter and Eliza Hall Institute, Peter MacCallum Cancer Institute and Murdoch Children's Research Institute. The Women in Science Parkville Precinct (WISPP) collaboration aims to tackle the broader issues limiting the progression of women in science. Even with our proactive initiative obtained after reading Dr White's report, it will be imperative that the institute reviews the success of these actions on a regular basis to truly see if we have been able to improve some of the obstacles limiting career progression, especially for women, highlighted by researchers in the report.

Geoffrey A Donnan AO MBBS MD FRACP FRCP (Edin.)
Director, The Florey Institute of Neuroscience & Mental Health
11 June 2014

APPENDIX
Interview Schedule

- How long have you worked at the Florey Neuroscience Institutes (FNI)?
- Could you please describe your role?
- What factors or people were most supportive in getting into your current position? And what factors have been less supportive?
- What gives you most satisfaction in your job?
- Is the current position your preferred career path in science? If not, what internal and external factors have impacted on your career? How important is financial reward?
- Do you consider that FNI provides optimum career opportunities for all staff?
- [For PhD students only] Has your perspective of your career in science changed since first year?
- [For early to mid-career postdocs only] Do you wish to become an independent researcher? If so, what do you see as the opportunities for and the barriers to doing so at FNI?
- Have you had any disruption to your career due to maternity/paternity leave, illness or the like? If so, has this impacted on career progression? And what have been the financial implications? If not, what is your perception of how this could affect your career?

- Would part-time career opportunities be useful for you?
- In your experience at FNI has gender been a factor in career progression and, if so, could you please discuss?
- What do you think are barriers to promotion for women and men at FNI?
- Are there ways in which promotion criteria could be altered to ensure greater equity in the promotions process?
- What role do different research networks play in women's and men's promotion at FNI and do women have the same opportunity as male colleagues to be introduced to these networks?
- What mentoring programs have been available to you at FNI? Have they been useful? If so, in what ways?
- What strategies do you think need to be implemented at FNI to ensure that women undertaking science PhDs and in the early career research phase have the same mentoring and support as their male colleagues?
- What initiatives do you think could be implemented to improve career progression and the position of women at FNI?

Questions for PIs/Lab Heads Only
- What are the challenges of your role at FNI?
- What do you consider is your role in developing career paths for your staff?
- Do you experience any conflict regarding your own career versus promoting younger members in your lab?
- What do you see as factors that impact on the ability of junior members to succeed?

Acknowledgements

This book has been a collaborative endeavour between the Equality in Science Committee at the Florey Institute of Neuroscience and Mental Health and the Principal Investigator.

I would like to thank the Institute's Director, Professor Geoff Donnan, who endorsed the project, as well as all members of the Equality in Science Committee. Special thanks go to Dr Jhodie Duncan, Florey Project Coordinator, Dr Jee Hyun Kim, Dr Emma Burrows, Astrid Sweres and Rodi Neri, Group Human Resources Manager, for their unfailing assistance. Thanks also to Professor John McDonald, Executive Dean of the Faculty of Education and Arts, Federation University Australia, for his support and to Jim Murray for editing the manuscript.

The forty researchers who agreed to be interviewed for this book provided a fascinating insight into the careers of Australian research scientists, and I would like to thank them for sharing their passion for science.

Dr Kate White
Principal Investigator
Adjunct Associate Professor
Faculty of Education and Arts
Federation University Australia
30 July 2014

References

Acker, S. Webber, M. and Smyth, E. (2010). Discipline And Publish? Early Career Faculty Meet Accountability Governance, New Managerialism and (Maybe) Gender Equity, paper presented to the Gender, Work and Organisation Conference, Keele University, 21–3 June.

Ackers, L. (2010). 'Internationalisation and Equality: The Contribution of Short Stay Mobility to Progression in Science Careers', *Recherches Sociologiques et Anthropologiques*, 40, 1, 83–103.

Asmar, C. (1999). 'Is There a Gendered Agenda in Academia? The Research Experience of Female and Male PhD Graduates in Australian Universities', *Higher Education*, 38, 255–73.

Association for Women in Science (AWIS) (2012). *The Work-Life Integration Overload* (AWIS: Alexandria, VA).

Association for Women in Science (NZ) (2011). *Women in Science: A 2011 Snapshot*. Viewed 7 December 2012, <www.awis.org.nz>

Australian Bureau of Statistics (ABS) (2013). *Australian Social Trends* (ABS: Canberra).

Australian Council of Learned Academies (ACOLA) (2012). *Career Support for Researchers: Understanding Needs and Developing a Best Practice Approach* (ACOLA: Canberra).

Bagilhole, B. and White, K. (eds.) (2013). *Generation and Gender in Academia* (Palgrave Macmillan: Basingstoke).

Bagilhole, B. and White, K. (eds.) (2011). *Gender, Power and Management: A Cross Cultural Analysis of Higher Education* (Palgrave Macmillan: Basingstoke).

Barakat, S. (2014). 'Women in Academia: Different Views of Success', *University World News Global Edition*, Issue 316, 18 April.

Barinaga, M. (1992). 'The Pipeline Is Leaking', *Science*, 255, 1366–7.

Barjak, F. and Robinson, S. (2008). 'International Collaboration, Mobility and Team Diversity in the Life Sciences: Impact on Research Performance', *Social Geography*, 3, 23–36.

Barrett, L. and Barrett, P. (2011). 'Women and Academic Workloads: Career Slow Lane or Cul-de-sac?' *Higher Education*, 61, 2, 141–155.

Bazeley, P., Kemp, L., Stevens, K., Asmar, C., Gribich, C., Marsh, H. and Bhathal, R. (1996). *Waiting in the Wings: A Study of Early Career Academic Researchers in Australia* (AGPS: Canberra).

Beaufays, S. and Kegen, N. (2012). Informal Networks in Science: Differences between Perceived and Realised Embeddedness of Female and Male Researchers, paper presented to the 7th European Conference on Gender Equality in Higher Education, Bergen, 29–31 August.

Bell, S. (2009). *Women in Science in Australia: Maximising Productivity, Diversity and Innovation* (FASTS: Canberra).

Benschop, Y. and Brouns, M. (2003). 'Crumbling Ivory Tower: Academic Organising and Its Gender Effects', *Gender, Work and Organisation*, 10, 2, 194–212.

Benschop, Y. (2009). 'The Micro-Politics of Gendering in Networking', *Gender, Work and Organisation*, 16, 2, 217–37.

Benavente, B. and González Ramos, A. (2014). Beyond Quality and Scientific Merit: A Feminist Exploration of the Scientific Career, paper presented to the Gender, Work and Organisation Conference, Keele University, 24–6 June.

Billiards, S. (2014). 'A Word on Women in Health and Medical Research—From the NHMRC', *Women in Science Australia*, 13 July. Viewed 15 July 2014,<http://womeninscienceaust.org/2014/07/13/a-word-on-women-in-health-and-medical-research-from-the-nhmrc/>

Birch, L. (2011). Telling Stories: A Thematic Narrative Analysis of Eight Women's PhD Experiences, PhD thesis, Victoria University.

Blackmore, P. and Kandiko, C. (2011). 'Motivation in Academic Life: A Prestige Economy', *Research in Post-Compulsory Education*, 16, 4, 399–411. doi:10.1080/13596748.2011.626971.

Bonetta, L. (2010). 'Reaching Gender Equity in Science: The Importance of Role Models and Mentors', *Science Careers*, 889–95. Viewed 10 February 2012, <sciencecareers.sciencemag.org/career...issues/.../science.opms.r1000084>.

Bradley, D. (2008). *Review of Australian Higher Education, Final Report* (Department of Education, Employment and Workplace Relations: Canberra).

Browning, L., Thompson, K. and Dawson, D. (2012). Gender Equality in Research Leadership, paper presented to 7th European Conference on Gender Equality in Higher Education, Bergen, 29–31 August.

Caprile, M. (coord.) (2012). *Meta-analysis of Gender and Science Research, Synthesis Report* (European Commission: Brussels).

Carral, M., Fusilier, B. and Murgia, A. (2014). Young Researchers Hanging in the Balance: The Experience of Precariousness between Professional and Private Lives, paper presented to Gender, Work and Organisation conference, Keele University, 24–6 June.

Carvalho, M. (2012). Towards the Definition/Perspectives on the Role of the Universities of Applied Sciences in the European Space, keynote address to the Second Conference, Universities of Applied Science Network, Braganza, 29 September.

Ceci, S. and Williams, W. (2011). 'Understanding Current Causes of Women's Underrepresentation in Science', *PNAS* 108, 8, 3157–62.

Ceci, S., Williams, W. and Barnett, S. (2009). 'Women's Underrepresentation in Science: Sociocultural and Biological Considerations', *Psychological Bulletin*, 135, 2, 218–61.

Charlesworth, M., Farrall, L., Stokes, T. and Turnbull, D. (1989). *Life Among the Scientists: An Anthropological Study of an Australian Scientific Community* (Oxford University Press: Melbourne).

Cidlinská, K. and Linková, M. (2013). Economy of Promise Failed: Career Plans of Women and Men at the Beginning of the Academic Path, paper presented to the Swiss Sociological Association Conference, Berne, 26–8 June.

Committee for Economic Development of Australia (CEDA) (2013). *Women in Leadership: Understanding the Gender Gap* (CEDA: Melbourne).

Corley, E. (2005). 'How Do Career Strategies, Gender and Work Environment Affect Faculty Productivity Levels in University Based Science Centres?' *Review of Policy Research*, 22, 5, 637–55.

Cory, S. (2011). 'So Seriously, Why Aren't There More Women in Science?', *The Conversation*, 24 August.

Cox, D. (2008). *Evidence on the Main Factors Inhibiting Mobility and Career Development of Researchers. Final Report to the European Commission* (Research Directorate General: Brussels).

Daphne Jackson Fellowships (2013). Viewed 3 March 2013, <www.daphnejackson.org>.

Dever, M., Boreham, P., Haynes, M., Kubler, M., Laffan, W., Behrens, K. and Western, M. (2008). *Gender Differences in Early Post-PhD Employment in Australian Universities: The Influence of the PhD Experience on Women's Academic Career, Final Report* (University of Queensland Social Research Centre: Brisbane).

De Vries, J. and Webb, C (2005). Gender in Mentoring: A Focus on the Mentor: Evaluating 10 Years of a Mentoring Programme for Women, paper presented to the Fourth European Conference on Gender Equality in Higher Education, Oxford, August.

Dubach, P., Graf, I., Stutz, H. and Gardiol, L. (2012). Dual-career Couples at Swiss Universities, paper presented to the Seventh European Conference on Gender Equality in Higher Education, Bergen, 29–31 August.

Ely, R. and Meyerson, D. (2000). 'Advancing Gender Equity in Organisations: The Challenge and Importance of Maintaining a Gender Narrative', *Organisation*, 7, 589–608.

Ely, R. and Padavic, I. (2007). 'A Feminist Analysis of Organisational Research on Sex Differences', *Academy of Management Review*, 32, 4, 1121–43.

Etzkowitz, H. and Kemelger, C. (2001). 'Gender Equality in Science: A Universal Condition?' *Minerva*, 39, 2, 239–57.

Etzkowitz, H., Kemelgor, C. and Uzzi, B. (1994). 'Barriers to Women in Academic Science and Engineering', in W. Pearson Jr. and I. Fechter (eds.), *Who Will Do Science? Educating the Next Generation* (Johns Hopkins University Press: Baltimore).

Euronews, 16 July 2012. Viewed 12 October 2013, <http://euronews.com/2012/07/16/erwin-neher>.

European Commission (2005a). *Research Headlines*, 12 September.

European Commission (2005b). *Women in Science: Excellence and Innovation: Gender Equality in Science* (European Commission: Brussels).

European Union (2012). *Structural Change in Research Institutions: Enhancing Excellence, Gender Equality and Efficiency in Research and Innovation* (Luxembourg: European Union).

Evans-Galea, M. (2012). Today's Early-mid Career Researcher Faces Many Challenges: Let's Discuss Some Possible Solutions, presentation to Florey National Institute, 1 May.

Faltholm, Y. and Abrahamsson, L. (2010). 'I Prefer not to be Called a Woman Entrepreneur'—Gendered Global and Local Discourses of Academic Entrepreneurship, paper presented to the Gender, Work and Organisation Conference, Keele University, 21–3 June.

Fitzgerald, T. and Wilkinson, J. (2010). *Travelling Towards a Mirage? Gender, Leadership and Higher Education* (Post Pressed: Mt Gravatt, Qld).

Florey Postdoctoral Association (2011). Survey of Florey National Institute Postdoctoral Researchers (FPA: Melbourne).

Florey Institute (2009). Women in Science—Activities in 2009.

Fouad, N. and Singh, R. (2011). *Stemming the Tide: Why Women Leave Engineering* (University of Wisconsin-Milwaukee: Wisconsin).

Fox, M. (2005). 'Gender, Family Characteristics, and Publication Productivity among Scientists', *Social Studies of Science* 35, 1, 131–50.

Fox, M. and Colletralla, C. (2006). 'Participation, Performance, and Advancement of Women in Academic Science and Engineering: What Is at Issue and Why', *Journal of Technology Transfer*, 31, 377–86.

Gender and Education Association (GEA) (2010). 'Representations of Women Scientists in Online Media', viewed 12 May 2014, <http://genderandeducation.com/issues/research-launch-women-scientists/comment-page-1>

genSET (2011). *Public Consultation on the Future of Gender and Innovation in Europe, Summary Report.* Viewed 10 November 2011 <www.genderinscience.org/.../EGS-Consultation-SummaryReport-Oct11-b-small.pdf>.

genSET (2010). *Recommendations for Action on the Gender Dimension in Science* (Portia: London).

Giles, M., Ski, C. and Vrdoljak, D. (2009). 'Career Pathways of Science, Engineering and Technology Research Postgraduates', *Australian Journal of Education*, 53, 1, 69–86.

González Ramos, A., Ibanez, M., Benavente, B. and Prieto, L. (2013). Excellence in Science: Is It Fair Play? paper presented to the Equality, Diversity and Inclusion Conference, Athens, 3–5 July.

González Ramos, A. and Vergés, N. (2012). 'International Mobility of Women in Science and Technology Careers: Shaping Plans for Personal and Professional Purposes', *Gender, Place and Culture*. 10.1080/0966369X.2012.701198.

Graves, N., Barnett, A. and Clarke, P. (2011). 'Funding Grant Proposals for Scientific Research: Retrospective Analysis of Scores by Members of Grant Review Panel', *British Medical Journal*, 343, 1–8.

Griffin, G. (2004). 'Tackling Gender Bias in the Measurement of Scientific Excellence: Combating Disciplinary Containment', *Gender and Excellence in the Making* (Office for Official Publications of the European Commission: Luxembourg), 127–33.

Grummell, B. Lynch, K. and Devine, D. (2009). 'Appointing Senior Managers in Education: Homosociability, Local Logics and Authenticity in the Selection Process', *Educational Management. Administration and Leadership*, 37, 3, 329–49.

Harman, G. (2010). 'Australian University Research Commercialisation: Perceptions of Technology Transfer Specialists and Science and Technology Academics', *Journal of Higher Education Policy and Management*, 32, 1, 69–83.

Hartley, N. and Dobele, A. (2009). ' Feather in the Nest: Establishing a Supportive Environment for Women Researchers', *The Australian Educational Researcher*, 36, 1, 43–58.

Hatchell, H. and Aveling, N. (2008). Gendered Disappearing Acts: Women's Doctoral Experiences in the Science Workplace, paper presented to the Australian Association for Research in Education Conference, Brisbane, 30 November—4 December.

Herbert, D., Coveney, J., Clarke, P., Graves, N. and Barnett, S. (2014). 'The Impact of Funding Deadlines on Personal Workloads, Stress and Family Relationships: A Qualitative Study of Australian Researchers', *British Medical Journal Open*, 4. Viewed 28 June 2014 <:e004462 doi:10.1136/bmjopen-2013-004462>.

Hewlett, S. and Buck Luce, C. (2006).'Extreme Jobs: The Dangerous Allure of the 70-Hour Week', *Harvard Business Review*, December, 49–59.

Holmquist, C. and Sundin, E. (2010). 'The suicide of the social sciences: Cause and effect', *Innovation—The European Journal of Social Sciences Research*, 23, 1, 13–23.

Hoppel, D. (2012). Ascent through Mentoring, paper presented to the Seventh European Conference on Gender Equality in Higher Education, Bergen, 29–31 August.

House of Commons Science and Technology Committee (HoC) (2014). *Women in Scientific Careers* (House of Commons: London).

Husu. L. (2004). 'Gate-keeping, Gender Equality and Scientific Excellence', in *Gender and Excellence in the Making* (Office for Official Publications of the European Commission: Luxembourg), 69–76.

Husu, L. and Koskinen, P. (2010). 'Gendering Excellence in Technological Research: A Comparative European Perspective', *Journal of Technology, Management & Innovation*, 5, 1, 127–39.

Ibarra, H., Carter, N. and Silva, C. (2010). 'Why Men Still Get More Promotions Than Women', *Harvard Business Review*, September, 88, 9, 80–5.

Ibarra, H., and Hunter, M. (2007).'How Leaders Create and Use Networks', *Harvard Business Review*, January, 81, 1, 40–7.

Jacobson, K. (2013). 'Six Steps to Fairer Funding for Female Scientists', *University World News Global Edition*, Issue 293, 25 October.

Jochimsen, M. and Muhlenbruch, B. (2009).'Gender Balance as a Precondition—Requirements for a Stronger Scientific Culture in the European Research Area'. In A. Lipinsky (ed.). *Encouragement to Advance— Supporting Women in European Science Careers* (Kleine Verlag: Bielefeld), 11–24.

Kanter, R.M. (1977). *Men and Women of the Corporation* (Basic Books: New York).

KPMG (2010). *Review of the Equal Opportunity for Women in the Workplace Act 1999 Consultation Report* (EOWA: Canberra).

Kretschmer, H. and Kretschmer, T. (2013). 'Gender Bias and Explanation Models for the Phenomenon of Women's Discriminations in Research Careers', *Scientometrics*, 97, 25–36.

Lane, N. (1999). 'Why Are There So Few Women in Science?', *Nature*, 9 September.

Leden, A., Bornmann, L., Gannon, F. and Wallon, G. (2007). 'A Persistent Problem: Traditional Gender Roles Hold Back Female Scientists', *EMBO Reports*, 8, 11, 982–7.

Lee, A., Dennis, C. and Campbell, P. (2007). '*Nature*'s Guide for Mentors', *Nature*, 14 June 447, 791–97.

Lind, I. and Lother, A. (2005). 'Gender Differences in Science Careers and Interventions for Women in Higher Education in Germany'. In V. Maione (ed.). *Gender Equality in Higher Education* (Franco Angela: Milan), 192–213.

Massachusetts Institute of Technology (1999). *A Study on the Status of Women Faculty in Science at MIT*. Viewed 25 June 2010, <http://web.mit.edu/faculty/reports/overview.html>.

Mavriplis, C., Heller, R., Beil, C., Dam, K., Yassinskaya, N., Shaw, M. and Sorensen, C. (2010). 'Mind the Gap: Women in STEM Career Breaks', *Journal of Technology, Management and Innovation*, 5, 1, 140–51.

McKeon Review (2013). *Strategic Review of Health and Medical Research* (Commonwealth of Australia: Canberra).

McNally, G. (2010). 'Scholarly Productivity, Impact, and Quality among Academic Psychologists at Group of Eight Universities', *Australian Journal of Psychology*, 62, 4, 204–15.

Mentornet (2014). Viewed 13 May 2014, <www.mentornet.com>.

Miles, M. and Huberman, M. (1994). *Qualitative Data Analysis*, Second ed. (Sage: London).

Minerva Fem Net (2012). Viewed 3 October 2012, <www.mpibpfrankfurt.mpg.de/misc>.

Monash University (2012). Viewed 6 September 2013, <http://monash.edu/equity-diversity/women/index.html>.

Monash University (2013). Viewed 6 September 2013, <http://monash.edu/equity-diversity/social-inclusion/eofw/gender-equity-grants.htm>.

Moir, J. (2006). Tipping the Scales: Talking about Women in Science and Work-Life Balance, paper presented at Science Policies Meet Reality: Gender, Women and Youth in Science in Central and Eastern Europe CEC-WYS conference, Prague, 1–2 December.

Morley, L. (2014). 'Lost Leaders: Women in the Global Academy', *Higher Education Research and Development*, 33, 1, 114–28.

Morley, L. (2013). *Women and Higher Education Leadership: Absences and Aspirations* (Leadership Foundation: London).

Morley, L. (1994). 'Glass Ceiling or Iron Cage: Women in UK Academia', *Gender, Work and Organisation*, 1, 4, 194–204.

Murray, F, and Graham, L. (2006). 'Buying Science and Selling Science: Gender Differences in the Market for Commercial Science, Industrial and Corporate Change', *Special Issue of Technology Transfer*, 16, 4, 657–89.

Nature Neuroscience (editorial) (2010). 'Wanted: Women in Research', 11, 3, 267.

Nature News, 2 June 2014.

O'Brien, K. and Hopgood, K. (2011). 'Part-time Balance', *Nature*, 479, 257–8.

O'Connor, P. (2011). 'Where Do Women Fit into University Senior Management? An Analytical Typology of Cross-National Organisational Cultures'. In B. Bagilhole and K. White (eds.), *Gender, Power and Management: A Cross Cultural Analysis of Higher Education* (Palgrave Macmillan: Basingstoke), 168–91.

Palacin, F., González Ramos, A. and Muñoz Márquez, M. (2013). Myths and Realities of Women Doing Science: The Inclusion of Women Scientists in a Male Scientific Mainstream, paper presented to Equality, Diversity and Inclusion conference, Athens, 3–5 July.

Peterson, H. (2014). 'An Academic "Glass Cliff"? Exploring the Increase of Women in Swedish Higher Education Management', *Athens Journal of Education* (February).

(2009). *Powering Ideas: An Innovation Agenda for the 21st Century* (Commonwealth of Australia: Canberra).

Research & Innovation (2005).

Reskin, B. and Roos, P. (1990). *Job Queues, Gender Queues, Explaining Women's In-Roads into Male Occupations* (Temple University Press: Philadelphia, AP).

Riordan, S. (2011). 'Paths to Success in Senior Management'. In B. Bagilhole and K. White (eds.), *Gender, Power and Management: A Cross-cultural Analysis of Higher Education* (Palgrave Macmillan: Basingstoke), 110–39.

Russo, G. (2010). 'For Love and Money', *Nature*, 465, 24 June, 1104–7.

Sabatier, M. Carrere, M. and Mangematin, V. (2006). 'Profiles of Academic Activities and Careers: Does Gender Matter? An Analysis Based on French Life Scientist CVs', *Journal of Technology Transfer*, 31, 311–24.

Sagebiel, F. (2013). Network Awareness and Successful Leadership in Science and Engineering, paper presented to the Swiss Sociological Association Conference, Berne, 26–28 June.

Sagebiel, F., Hendrix, U. and Schrettenbrunner, C. (2011). Women Engineers and Scientists at the Top as Change Agents? paper presented to Gender Renewals? Gender Work and Organisation International Workshop Series, VU University Amsterdam, 22–4 June.

Sagebiel, F., Hendrix, U. and Schrettenbrunner, C. (2010). How Women Scientists at the Top Change Organisational Cultures, paper presented to the ISA World Congress of Sociology, Gothenburg, 11–7 July.

Science Careers Blog, 23 August 2012.

Sekreta, E. (2006). 'Sexual Harassment, Misconduct, and the Atmosphere of the Laboratory: The Legal and Professional Challenges Faced by Women Physical Science Researchers at Educational Institutions', *Duke Journal of Gender Law and Policy*, 13, 115–37.

Shah, R. (2011). 'Working with Five Generations in the Workplace', *Forbes*, 20 April.

Smaglik, P. (2011). *Naturejobs*, 469, 121–3.

Stark, L. (2008). 'Exposing Hostile Work Environments for Female Students in Academic Science Laboratories: The *McDonnell Douglas* Burden-Shifting Framework as a Paradigm for Analysing the "Women in Science" Problem', *Harvard Journal of Law & Gender*, 31, 101–168.

The Conversation (2014). Viewed 10 April 2014, <http://theconversation.edu.au/so-seriously-why-arent-there-more-women-in-science-2917>.

The Conversation (2011). Viewed 5 April 2012, <http://theconversation.edu.au/a-breakthrough-for-women-in-science-764>.

UNESCO (2012). *World Atlas of Gender Equality in Education*. Viewed 10 March 2014, <www.uis.unesco.org/Education/Documents/unesco-world-atlas-gender-education-2-12.pdf>.

van den Brink, M., Benschop, Y. and Jansen, W. (2010). 'Transparency in Academic Recruitment: A Problematic Tool for Gender Equality?' *Organisation Studies*, 31, 11, 1459–83.

van den Brink, M. (2014). Sustainable Change Towards Diversity: The Role of Sponsoring, paper presented to the Gender, Work and Organisation conference, Keele University, 24–6 June.

van den Brink, M. (2009). Behind the Scenes of Science: Gender Practices in the Recruitment and Selection of Professors in the Netherlands, PhD thesis, University of Nijmegen.

Van Noorden, R. (2014). 'Scientists and the Social Network', *Nature*, 512, 126–9.

Vlasak, O. (2012). Integrating the Gender Dimension in Research and Innovation Content, presentation to Second European Gender Summit, Brussels, 29–30 November.

von Alemann, A. and Beaufays, S. (2014). Theorizing Gender (In)equality in the Work-family Balance: Fathers' Careers and Care Work—A Case of Feminised Demotion in Organisations?, paper presented to the Gender, Work and Organization Conference, Keele University, 24–6 June.

Walter and Eliza Hall Institute (WEHI) (2013). Viewed 23 August 2013, <www.wehi.edu.au/about_us/gender_equity/>.

Walter and Eliza Hall Institute (WEHI) (2012). Viewed 3 June 2012, <www.wehi.edu.au/site/latest_news/new_initiatives_support_women_scientists>.

Wenneras, C. and Wold, A. (1997). 'Nepotism and Sexism in Peer Review', *Nature*, 387, 341–3.

West, C. and Zimmerman, D. (1987). 'Doing Gender', *Gender and Society*, 1, 2, 125–51.

White, K. and Bagilhole, B. (2013). 'Continuity and change in academic careers'. In B. Bagilhole and K. White (eds.), *Generation and Gender in Academia* (Palgrave Macmillan: Basingstoke), 169–95.

White, K. (2011). 'Legislative Frameworks for EO'. In B. Bagilhole and K. White (eds.), *Gender, Power and Management: A Cross-cultural Analysis of Higher Education* (Palgrave Macmillan: Basingstoke), 20–49.

Williams, A. (2010). Women's Access to and Participation in Science and Technology, presentation to UN Commission on the Status of Women, fifty-fourth session, New York, 1–12 March.

Wilson, R. (2012). 'Scholarly Publishing's Gender Gap: Women Cluster in Certain Fields, According to a Study Of Millions of Journal Articles, While Men Get More Credit', *The Chronicle of Higher Education*. Viewed 10 March 2014, <http://chronicle.com/article/The-Hard-Numbers-Behind/135236/?cid=wb&utm_source=wb&utm_medium=en>.

Wilson-Kovacs, D., Ryan, M. and Haslam, S. (2006). 'The Glass-cliff: Women's Career Paths in the UK Private IT Sector', *Equal Opportunities International*, 25, 8, 674–87.

Witz, A. and Savage, M. (1992). *Gender and Bureaucracy* (Blackwell: Oxford).

Wroblewski, A. (2010). Barriers to Women on Their Way into Top Positions in Austrian Universities: How Gender Biased are Application Procedures for University Professors?, paper presented to ISA World Congress of Sociology, Gothenburg, 11–17 July.

Yu, X. and Shauman, K. (2003). *Women in Science: Career Processes and Outcomes* (Harvard University Press: Cambridge, MA).

Zippel, K. (2012). Gender in the Globalizing Academic World, keynote address to the seventh European Conference on Gender Equality in Higher Education, Bergen, 27–9 August.

Index

Acker, S., Webber, M. and Smyth, E. 17, 179
Ackers, L. 17–18, 80–2, 160, 179
ACOLA 4, 89, 129, 146, 179
Asmar, C. 13, 102, 179
Association of Women in Science (AWIS) (US) 6–7, 70, 120–1, 127, 134, 139, 148, 164–5, 179
AWIS (NZ) 33
Austin Hospital *see* Florey Institute of Neurosciences and Mental Health
Australia 141, 166
Australian Bureau of Statistics 19, 121, 179
Australian government 1, 141
   Fairwork legislation 105
   Innovation Agenda 1, 141, 166, 185
Australian Postgraduate Awards (APAs) 9, 37
Australian Research Council 8, 63, 114
   centres and networks 19
   Laureat Fellowships 20

Bagilhole, B. 23
Bagilhole, B. and White, K. 23, 101, 159, 179
Barakat, S. 161, 179
Barinaga, M. 17, 179
Barjak, F. and Robinson, S. 17, 180
Barrett, L. and Barrett, P. 101, 180
Bazeley, P., Kemp, L., Steven, K., Asmar, C., Gribiche, C., Marsh, H. and Bhathal, R. 57, 180
Beaufays, S. and Kegen, N. 69, 180
Bell, S. 2, 7, 13–14, 16, 20, 33–4, 102, 167, 180
Benavente, B. and González Ramos, A. 163, 180
Benschop, Y. 18, 91, 160, 180
Benschop, Y. and Brouns, M. 160, 180
Billiards, S. 166, 180
Birch, L. 13, 34, 180
Blackmore, P. and Kandiko, C. 50, 180
Bonetta, L. 18, 70, 180
Boston 63
Bradley Report 32, 180

Browning, L., Thompson, K. and Dawson, D. 159, 180
Burrows, E. 178

Caprile, M. 2, 15–16, 33, 81, 94, 98, 100–1, 158, 161–3, 167, 180
careers/career paths (in research science) 25–7
  absence of women 95–6
  attrition of women 2, 7, 10, 32, 54–5, 116
  barriers for women 2–3
  becoming an independent researcher 50–2, 60–2
  building career paths 13, 54–68, 71ff
  choice (and gender) 15, 98–100, 109–18
  collaboration 9, 17, 19–21, 23, 46, 63–4, 72, 74–5, 77–8, 81–2, 108–18, 120, 154, 160, 175
  communicating science 46
  competition 47, 99
  early career researchers 13, 23, 25, 37, 41, 50, 52, 56ff, 67, 71, 77, 80, 89–91, 93, 101, 148, 156–7, 159–61, 169, 171, 178
  financial reward 48–50
  flexibility 52–3
  gatekeeping 16–17
  gender bias 13–14, 19, 93–118
  gender discrimination 93–118
  gendering of 93–118
  generational change and *see* generational change
  impact of funding on *see* NHMRC
  initiatives to improve women's careers 20–1
  intellectual freedom 52–3
  interrupted careers 15–16, 108–11, 125–40
  job satisfaction 43–53
  lack of transparency in 42, 65, 91, 159, 163, 169–70, 174
  maternity leave 4, 21, 107, 109–11, 135, 143, 146, 148–55, 163, 166–8
  mobility and *see* mobility
  passion for science 44–5
  poor salaries 49, 53, 58, 120, 145, 148
  'problem is women' 96–7
  satisfaction levels, key factors 48–53
  securing external funding 3, 27, 50, 62, 67, 108
  support of supervisors 47, 61, 72, 108–11, 114–16, 163
  working part-time 107–11, 114–15, 127–140
career stages 26–7, 56–67, 143–5
  PhD students 26, 56–7, 67
  early, mid career 26, 56–8, 67
  postdocs 26–7, 56–8, 67
  research officers 27, 56–8, 67
  senior research officers 27, 56, 60–2, 67
  research fellows 27, 56, 62–8
  senior research fellows 27, 56, 62–7
Carral, M., Fusilier, B. and Murgia, A. 101, 105, 180
Carvalho, M. 142, 180
Ceci, S., Williams, W. and Barnett, S. 20, 181
Ceci, S. and Williams, W. 13, 15–16, 20, 115, 167, 181
CEDA 14, 181
Charlesworth, M., Farrall, L., Stokes, T. and Turnbull, D. 6, 47, 50, 55, 181
Cidlinská, K. and Linková, M. 163, 181
Commonwealth Scientific and Research Organisation (CSIRO) 19–20
Cooperative Research Centres 19
Corley, E. 15, 181
Cory, S. 90, 158, 167, 181
Cox, D. 15, 181

Daphne Jackson Fellowship (UK) 171, 181

Index 189

Department of Florey Neurosciences (University of Melbourne) 8
De Vries, J. and Webb, C. 96, 181
Dever, M., Boreham, P., Haynes, M., Kubler, M., Laffan, W., Behrens, K., Western, M. and Dickson, G. 13, 181
Donnan, G. AO 8, 23, 93, 173–5
Dubach, P., Graf, I., Stutz, H. and Gardiol, L. 80, 115, 181
Duncan, J. 178

Ely, R. and Meyerson, D. 98, 160, 181
Ely, R. and Padavic, I. 12, 181
Etzkowitz, H. and Kemelger, C. 13, 17, 181
Etzkowitz, H., Kemelger, C. and Uzzi, B. 12–13, 101, 117, 181
Euronews 119, 182
Europe 32, 70, 76, 79, 119, 142, 160
European Commission 12, 166, 182
European Union 10, 141–2
Evans-Galea, M. 41, 182

Faltholm, Y. and Abrahamsson, L. 18, 182
Federation University Australia (formerly University of Ballarat) 23, 29, 178
   Human Research Ethics Committee 23, 29
Fitzgerald, T. and Wilkinson, J. 98, 182
Florey Institute of Neuroscience and Mental Health 7–10, 20–31, 93–118
   analysis of workforce 32–9
   Austin hospital campus 7–8, 24, 26, 36–7, 42, 108, 125, 156, 162, 165
   Brain Research Institute 8, 36
   clinical versus research 108, 125
   Ethics Committee 170
   Equality in Science Committee (EqIS) 10, 23, 26–7, 30, 71, 91, 93–4, 112, 168–9, 171, 173–4
   faculty 63
   Florey Fellowship 62–3
   funding 8–10, 30, 38–9, 51, 55, 58–9, 65–7
   increasing visibility of women in science 174
   mentoring program 83, 87, 174
   National Stroke Research Institute 36–7
   Occupational Health and Safety Committee 170
   Organisational culture 93–118, 163
   Parkville campus 7, 22, 24, 26–7, 34, 36–40, 42, 71, 77, 108, 116, 125, 149, 156–7, 162, 170, 175
   postgraduate student association 9, 71, 83
   Promotions Committee 36, 62, 170
   recruitment of researchers 51, 62, 107, 109–10, 112, 143
   research areas 7
   start-up packages *see also* funding 149ff
   strategies to keep women in science 119–55, 174–5
   targets to increase gender representation 169
   Women in Science group 10, 20, 22
   Workforce profile by gender: part-time, full-time, casual 7, 35–42
Florey, Lord Howard 7
Florey Neuroscience Institutes (FNI) 8, 22, 70, 176–7
Florey Postdoctoral Association (FPA) 182
   survey 9–10
Fouad, N. and Singh, R. 17–18, 182
Fox, M. 15, 182
Fox, M. and Colletralla, C. 15, 182
Fred P. Archer Fellowship 36, 96
funding *see also* NHMRC 3–7, 14, 103, 105–7
   assessment of output relative to opportunity 10, 14, 104, 106–7, 110–11, 128–30, 137–9, 148, 155

effectiveness of funding model 3, 5–6, 10–11, 21, 54, 65, 94–5, 105–7, 117–18, 126–30, 142, 147, 163, 165, 168
  inconsistency in 66
  part-time work and 4, 10–11, 20, 23, 93, 105, 107, 109, 114–15, 128–30, 132, 40, 143–7, 149, 154–5, 164–6, 168, 171–2, 175

Gender and Education Association (GEA) 158, 182
gender in science research *see also* careers 13–14, 19, 93–118
  homosociability/homophilius networks 18
Gender, Power and Management see Bagilhole and White
generational change *see also*
  academic careers 118–40, 175
  Baby Boomers 119–40, 164
  communication/technology and 122, 126
  compressed working week 131–6
  dual careers 28, 49, 79, 127, 130–1, 136, 138, 150, 165, 171–2
  family 119–40
  flexible work models 131–6
  Generation X (Gen X) 119–40, 164
  Generation Y (Gen Y) 119–40, 164
  impact of funding on 119–40
genSET 2–3, 12–14, 19–20, 79, 116, 182
Giles, M., Ski., C and Vrdoljack, D. 75, 101, 182
González Ramos, A., Ibanez, M., Benavente, B. and Prieno, L. 140, 182
González Ramos, A. and Vergés, N. 53, 75, 182
Google Scholar 126
Graves, N., Barnett, A. and Clarke, P. 3, 14, 107, 183
Griffin, G. 17, 183
Grummell, B. Lynch, K. and Devine, D. 18, 91, 102, 163, 183

Harman, G. 17, 183
Hartley, N. and Dobele, A. 15, 19, 183
Hatchell, H. and Aveling, N. 13, 183
Herbert, D., Coveney, J., Clarke, P., Graves, N. and Barnett, S. 4, 107, 165, 183
Hewlett, S. and Buck, Luce, C. 6, 183
Holmquist, C. and Sundin, E. 17, 183
Hoppel, D. 70, 183
House of Commons Committee Report (HoC) (UK) 1, 4, 18–19, 55, 60, 70, 78–9, 89, 92, 107, 112–14, 121, 127, 136, 139, 142–3, 145, 148, 151, 159, 161, 167, 183
Howard Florey Research Institute 7
Husu, L. 16, 18, 183
Husu, L. and Koskinen, P. 13, 17, 20, 183
Hyun Kim, J. 178

Ibarra, H., Carter, N. and Silva, C. 88, 92, 183
Ibarra, H. and Hunter, M. 69, 73, 184
Institute of Physics and Engineering in Medicine (UK) 145
interview schedule 27, 176–7

Jackson, G. 8
Jacobson, K. 4, 16, 184
Jochimsen, M. and Muhlenbruch, B. 166, 184

Kanter, R.M. 13, 184
KPMG 16, 184
Kretschmer, H. and Kretschmer, T. 94, 184

Lane, N. 15, 184
Leden, A., Bornmann, L., Gannon, F. and Wallon, G. 18, 184
Lee, A., Dennis, C. and Campbell, P. 83, 85, 184
Lind, I. and Lother, A. 20, 184
London 63

McDonald, J. 178
McKeon Report/Review 27, 34, 57–8, 61, 66, 96, 108, 184
  recommendations 1, 5, 7, 135, 164, 168
  terms of reference 4–5
McNally, G. 15, 94, 184
Massachusetts Institute of Technology (MIT) 16, 184
Masters, C. 8
Mavriplis, C., Heller, R., Beil, C., Dam, K., Yassinskaya, N., Shaw, M. and Sorensen, C. 15, 184
Max Planck Institute of Biophysics 70
Melbourne 2, 6–9, 47, 63, 124, 150, 156
Melbourne University Research Scholarships 9
Mendelsohn, F. 8
Mental Health Research Institute 8
mentoring *see also* careers 18–19, 69–70, 83–92, 157
  and career development 10, 69, 72, 74–6, 81–2, 84–6, 88, 92, 110, 140, 159–60, 162, 167–9, 171, 175
  and gender 89–92
  meta-analysis of programs 70
  sponsorship 88–9, 162
  supervisor as mentor 84–5
  value of 83–4
MentorNet 70
methodology *see* research design
Miles, M. and Huberman, M. 26, 184
Milgrom, N. 151
Minerva Fem-Net 70
Mitchell, H. AC 174
mobility *see also* careers/careers paths and networking 15, 17–18, 69–70, 75–82, 160–2
  benefits of international collaboration 75–82
  and family 17–18, 21, 28, 75–82, 92
  and gender 75–82

  at postdoc phase 75–82
  shift in attitude 75–82
  short-term travel 78–82
Moir, J. 5, 13, 15, 98, 100–1, 116, 159, 164, 185
Monash University
  advancing women in research grant program 150
  gender equity travel support grants 150
Morley, L. 19, 88, 96, 101, 142, 185
Morris, C. 10
Murdoch Children's Research Institute 175
Murray, F. and Graham, L. 13, 15, 17, 185
Murray, J. 178

National Health and Medical Research Council (NHMRC) *see also* funding 3–5, 8–10, 14, 19, 41, 51, 58, 63, 66, 78, 104–8, 110–14, 123, 126, 128–30, 135–6, 166, 171, 180
  Career Development Award (CDA) 10, 41, 66, 113
  Career Development Fellowship (CDF) 78
  Chief Knowledge and Development Officer 10
  facts book 41
  project grants 3–4, 66
  Senior Research Fellowship 66, 78
  strategies to retain women scientists 126
National Stroke Research Institute 8, 36–7
Nature 48–50, 52, 66, 130, 158, 184–7
Nature Neuroscience 15, 142, 185
Nature News 16, 185
Naturejobs 6, 54, 186
Nature's Guide for Mentors 83, 85, 184
Nature's salary and career survey *see* Russo, G.
Neher, E. 119, 182
Neri, R. 27, 178

networking 18, 69–82, 101, 152, 157
   and careers 56, 71, 114, 123
   and conferences 72–82
   gender differences 72–4
   role of supervisor in 72
   seminars 71, 159
   social functions 71, 73–4
   student association role in 71
   visiting speakers 71
   weekly seminars 71
New Zealand 33

O'Brien, K. and Hopgood, K. 20, 185
O'Connor, P. 97, 185
Office for Women (Australia) 19

Palacin, F., González Ramos, A. and Muñoz Márquez M. 46, 185
Parliament House (Canberra) 2
Peter McCallum Cancer Institute 175
Peterson, H. 14, 185
Powering Ideas 1, 166

recommendations 170–1
research design
   aim 24, 156
   expected outcome 29–30
   hypothesis 24, 171
   method 25–9
   objectives 24, 156–7
   research questions 25
   significance and innovation 30–1, 168–9
ResearchGate 126
Research and Innovation 48, 185
Reskin, B. and Roos, P. 14, 185
Riordan, S. 56, 58, 66, 159, 165, 185
Royal Academy of Engineers (UK) 136
Russell Group Equality Forum (UK) 121
Russo, G. 48–53, 185

Sabatier, M., Carrere, M. and Mangematin, V. 17–19, 157, 185
Sagebiel, F. 19, 73, 185

Sagebiel, F., Hendrix, U. and Schrettenbrunner, C. 18, 70, 157, 160, 185
Science Careers Blog 6, 186
Science, Technology, Engineering and Mathematics (STEM) 1, 13, 19, 70, 89, 112, 142–3, 159, 167, 184
scientific research
   authorship 9, 16–17, 72, 108–10, 143–5, 167
   changes in 2, 47, 102, 119
   collegiality 46–7
   competitive collaboration 47
   editorial boards 72
   gender analysis in 19, 43
   long hours work culture 127–31, 158, 165
   model of successful scientist 5, 10, 12–13, 94, 101, 117–18, 158
   picking winners 51, 62, 107, 109–10
   publication 2–4, 10, 14, 16–17, 45, 55, 66, 69, 72, 79, 94, 99, 108–11, 116, 130–1, 137–41, 144–5, 147, 154–5
   publication in high impact journals 17, 107, 111, 130
   track record 4, 20, 65–6, 94, 99, 106, 110, 170
Sekreta, E. 16, 186
Shah, R. 119, 186
Site, K. 37
Smaglik, P. 54, 186
South East Asia 141
sponsorship *see* mentoring
Stark, L. 16, 101–3, 186
supervisors *see* careers/career paths
Sweres, A. 178

The Conversation 14, 20–1, 181, 186

UNESCO 142, 186
United States 23, 76, 77, 79, 160, 181
United Kingdom 1, 19, 23, 60, 70, 78, 89, 107, 112–14, 121, 136, 142, 145, 148, 150, 159, 167, 171

Index 193

University of Manchester 145
University of Melbourne 7

van den Brink, M. 13, 16, 18, 33, 89, 162, 186
van den Brink, M., Benshop, Y. and Jansen, W. 12, 16–17, 186
Van Noorden, R. 126, 186
Vlasak, O. 142, 186
von Alemann, A. and Beaufays, S. 5, 13, 166, 186

Walter and Eliza Hall Institute (WEHI) 6, 21, 150, 175, 186
  Craven and Shearer Award 150
Wenneras, C. and Wold, A. 13, 187
West, C. and Zimmerman, D. 12, 187
White, K. 23, 105, 173, 175
White, K. and Bagilhole, B. 88, 187
Willetts, D. Rt Hon 60
Williams, A. 2, 187
Wilson, R. 17, 167, 187
Wilson-Kovacs, D., Ryan, M. and Haslam, A. 19, 187
Witz, A. and Savage, M. 163, 187
Women in Science Parkville Precinct (WISPP) 175

work–life balance *see also* careers and generational change 1, 70, 74–6, 90–1, 94ff, 100–1, 108–18, 121–40, 142ff, 167
  childcare 20–1, 114, 146
  family friendly meeting times/ working hours 15, 21, 151, 175
  family rooms in the workplace 21, 175
  flexible working hours/ arrangements 16, 21, 52, 105, 121, 127, 131–40, 144–6, 165, 168, 171, 175
  long hours work culture 5–6, 52, 55, 67–8, 123, 159
  new models 131–40, 151–5
  part-time work 4, 10–11, 13, 20, 22–3, 37–8, 42, 80, 93, 98, 106–7, 109, 114–15, 129–30, 132–5, 137–40, 143–149, 154–5, 164–8, 171
  support from supervisor 148–55
  transitioning back to work 20, 148–55
Wroblewski, A. 18, 157, 187

Yu, K. and Sharman, K. 13, 187

Zippel, K. 18, 75, 78, 80, 161, 187